SpringerBriefs in Food, Health, and Nutrition

Springer Briefs in Food, Health, and Nutrition present concise summaries of cutting edge research and practical applications across a wide range of topics related to the field of food science.

Editor-in-Chief
Richard W. Hartel
University of Wisconsin—Madison, USA

Associate Editors
J. Peter Clark, *Consultant to the Process Industries, USA*
John W. Finley, *Louisiana State University, USA*
David Rodriguez-Lazaro, *ITACyL, Spain*
David Topping, *CSIRO, Australia*

For further volumes:
http://www.springer.com/series/10203

Lucy J. Robertson

Giardia as a Foodborne Pathogen

Springer

Lucy J. Robertson
Institute for Food Safety & Infection Biology
Norwegian School of Veterinary Science
Oslo, Norway

ISBN 978-1-4614-7755-6   ISBN 978-1-4614-7756-3 (eBook)
DOI 10.1007/978-1-4614-7756-3
Springer New York Heidelberg Dordrecht London

Library of Congress Control Number: 2013940159

Printed on acid-free paper

Springer is part of Springer Science+Business Media (www.springer.com)

Contents

Contents

Chapter 1
Introduction to *Giardia duodenalis*:
The Parasite and the Disease

Giardia is a genus of protozoan parasites within the phylum Metamonada, the order Diplomonadida, and the family Hexamitidae. Six different species are considered to belong to the *Giardia* genus:

- *Giardia duodenalis* (syn. *Giardia lamblia*, *Giardia intestinalis*) that infects a broad range of mammalian hosts
- *Giardia agilis* that infects amphibians
- *Giardia muris* that infects mice
- *Giardia microti* that infects voles
- *Giardia ardeae* and *Giardia psittaci* that both infect birds

For the purposes of this Springer brief, we focus only upon *Giardia duodenalis*, which is of importance with respect to both public and veterinary health.

Host specificity and genetic differences have led to the suggestion that *Giardia duodenalis* is a species complex and, despite lack of morphological differences, should be redescribed as a number of different species, such as *Giardia enterica* (in humans and other primates, and some other mammals), *Giardia canis* (in dogs), *Giardia cati* (in cats) and *Giardia bovis* (in cattle and other hooved livestock) (Monis et al. 2009). Although there is some compelling evidence in favour of this proposed reclassification, it has not yet been widely accepted. Instead, the species is currently divided into a number of genetically distinct groups, known as Assemblages, and some of these Assemblages have been further subdivided into genotypes. The various Assemblages and genotypes are also characterised by particular host specificities; *G. duodenalis* in Assemblage A1 is probably the most important zoonotic genotype, A2 predominantly infects humans, but may also be zoonotic, while A3 is common among wild ungulates (Sprong et al. 2009; Beck et al. 2011). *Giardia* in Assemblage B appears to be more heterogenic (Wielinga et al. 2011), but is predominantly found in humans and can also be zoonotic. Nevertheless, the importance of giardiasis (also known as giardiosis) as a zoonosis remains unresolved, and it seems that the majority of data suggests that most *Giardia* infections in animals pose little or no risk to public health. In general, *Giardia* in

L.J. Robertson, *Giardia as a Foodborne Pathogen*, SpringerBriefs
in Food, Health, and Nutrition, DOI 10.1007/978-1-4614-7756-3_1,
© Lucy J. Robertson 2013

Assemblages C and D appear to exclusively infect canids, Assemblage E infects ruminants, Assemblage F infects cats and Assemblage H infects pinnipeds (Lasek-Nesselquist et al. 2010). Other non-morphological differences have also been demonstrated among the different Assemblages, including differences in pathology and growth dynamics (Bénéré et al. 2011, 2012) and *in vitro* drug susceptibilities (Bénéré et al. 2011).

G. duodenalis is generally considered to have a global distribution and is the most common intestinal parasite of humans, with over 2.5×10^8 cases annually (Lane and Lloyd 2002; Cook and Lim 2012). The most common intestinal protozoan parasite in all countries of the world, in developing countries, giardiasis is particularly common, and is especially predominant in preschool and school children, with the prevalence estimated to reach over 70 % in some populations (Cook and Lim 2012; Dib et al. 2008). In Africa, Asia and Latin America it has been estimated that symptomatic giardiasis occurs in around 200 million people, and that approximately 500,000 new cases are reported on an annual basis (Cacciò and Sprong 2011). In 2004, the common link of giardiasis with poverty encouraged its inclusion in the WHO "Neglected Diseases Initiative" (Savioli et al. 2006).

Within Europe, several countries collect data on cases of giardiasis and report them to the European Centre for Disease Control. Although the true incidence is almost certainly underestimated and reporting rates are likely to be influenced by factors other than the actual occurrence of infection, including laboratory capabilities, medical awareness of the disease and other national policies and idiosyncrasies, these do provide some useful comparative data. According to Cacciò and Sprong (2011), during 2006 the highest rates of infection in Europe were reported from Romania, followed by Estonia, followed by Bulgaria, followed by Sweden. The unexpectedly high rate of infection reported from Sweden is likely to reflect national competency in diagnosis as much as higher infection rates.

In animals, prevalence data are more sparse; although several studies regarding the prevalence of *Giardia* infection in individual animal species and regions have been published, extrapolation from these data to global estimates is not possible due to wide regional and population-specific variations, and also due to huge variation in study design regarding how the prevalence data have been collected. However, as humans appear to be the major source of infections for humans, and the available sub-genotyping data do not indicate the widespread occurrence of zoonotic transmission (Geurden and Olsen 2011), the prevalence of animal infections will not be considered in further detail here.

The life cycle of *G. duodenalis* is simple and direct, and comprises two morphologically distinct forms: the vegetative trophozoites (characteristically pear-shaped, 9–20 μm by 5–15 μm with two nuclei, eight flagella, linear axonemes, curved median bodies and a ventral adhesive disc) that inhabit the lumen of the small intestine—either free-swimming or attaching onto the enterocyte brush border of the mucosal surface by the adhesive disk—and the environmentally resistant cysts (ovoid, 8–18 μm by 7–10 μm).

Although trophozoites of *G. duodenalis* are generally considered to replicate only asexually, by simple binary fission, evidence suggests that genetic exchange probably does occur, although perhaps infrequently. Nevertheless, the mechanism of sexual reproduction remains unresolved, and the significance of sexual reproduction to the pathogenicity and epidemiology of *Giardia* is also unknown (Birky 2010). However, repeated binary fission results in the establishment of enormous number of trophozoites, and these can cover the mucosal surface of the intestine. Encystation from trophozoites to cysts occurs as the trophozoites move further down the intestine, and happens in response to changing mixtures of hydrogen ions, bile salts, proteases and other conditions. Although trophozoites might also sometimes be excreted in the faeces (especially when diarrhoea is particularly severe), cysts are considered as the transmission stage and are immediately infectious upon excretion; unless maintained in conditions of *in vitro* culture, trophozoites do not survive for long after excretion.

The *Giardia* cyst wall (between 0.3 and 0.5 μm in thickness) has a filamentous structure, containing carbohydrates and proteins in a ratio of 3:2 (w/w), with the carbohydrate moiety composed at least partially of a beta (1–3)-*N*-acetyl-D-galactopyranosamine homopolymer (Gerwig et al. 2002). Two distinct regions of the cyst wall have been identified, with the external region, which is probably most protective against environmental pressures, composed of bundles of fibrils between 7 and 22 nm in thickness, and connected to each other by short and thin filaments (Benchimol and De Souza 2011).

It has been suggested that the polysaccharide of the cyst wall forms ordered helices, or possibly multiple helical structures, with strong interchain interactions, and this structure ensures the robustness of the cyst wall, and thereby enables survival of the cysts for prolonged periods in damp environments (Gerwig et al. 2002).

Infection with *G. duodenalis* is initiated when a viable cyst is ingested by a susceptible host. This may be direct faecal-oral ingestion, or via a vehicle such as contaminated water or food. The infective dose is theoretically a single cyst; in early infection studies a dose of ten cysts was reported to result in infection (although not necessarily disease) in two out of two volunteers (Rendtorff 1954). It should be noted, however, that not all human-source isolates are equally infectious to all people (Nash et al. 1987). Exposure to factors such as gastric acid, pepsin and the alkaline environment of the small intestine triggers excystation of the cysts in the upper small intestine. Two trophozoites are released from each cyst, and infection is established as these trophozoites develop and replicate within the host intestine. Symptoms may begin around 1–2 weeks after infection, and this relatively long period between infection and disease is a confounder for identifying the source of the infection, particularly if a food vehicle is suspected. People are remarkably poor at remembering possible exposure routes or foods that they have encountered a week or so back, and it is also likely that potential food vehicles are likely either to have been consumed or disposed of before the infection is diagnosed (which, again, may be several days after the commencement of symptoms).

Human infection with *G. duodenalis* is generally associated with diarrhoea, which tends to be fatty and foul-smelling, but can also be either asymptomatic or responsible for a broad clinical spectrum, with symptoms ranging from acute to chronic (Robertson et al. 2010). Generally it is estimated that of individuals exposed to infective *Giardia* cysts, for 50 % no clinical symptoms are manifest, and the infection clears without treatment and may not even establish properly, for between 5 and 15 % infection establishes and cysts are shed, but no symptoms (or only very mild, intermittent symptoms) are noticed, while for the remaining group (between 35 and 40 % of those exposed) symptomatic infection establishes. Although giardiasis is not generally associated with mortality, estimates have been made of annual deaths associated with giardiasis (Nuñez and Robertson 2012). While in the UK (England and Wales), Adak et al. (2002) considered that no deaths occurred from giardiasis, three different estimates from the USA have suggested that a few deaths (between 1 and 10) may occur annually due to giardiasis (Mead et al. 1999; Frenzen 2004; Scallan et al. 2011). Indeed, Mead et al. (1999) estimated ten deaths due to giardiasis in the USA annually, of which one would be due to foodborne giardiasis. In situations in which other health and lifestyle factors, particularly those associated with poor nutrition and other infections, may occur concomitantly, infection with *Giardia* may play a synergistic role, perhaps leading to more severe symptomatology or even death.

Chronic giardiasis is usually associated with loose stools and/or diarrhoea, along with intestinal malabsorption, resulting in steatorrhoea, lactase deficiency and vitamin deficiencies. Potential mechanisms for this include disruption in epithelial transport and barrier dysfunction. Abdominal pain, bloating, nausea, vomiting, failure to thrive and anorexia are also all common symptoms, and may be accompanied by profound weight loss (loss of 10–20 % of original body weight). Up to 40 % of individuals that suffer symptomatic giardiasis also develop acquired lactose intolerance, which is manifest as an exacerbation in intestinal symptoms following ingestion of dairy products (Cantey et al. 2011). It should also be noted that nutrient malabsorption (particularly of fats, sugars, carbohydrates and vitamins) and alterations in the activity of enterocytes may occur, even in the absence of overt symptoms of giardiasis. This may result in hypoalbuminaemia and also deficiencies in vitamins A, B12 and folate. For individuals that are living in situations where dietary intake of vitamins may be limited, this malabsorption may mean the difference between deficiency and clinical disease.

Some cases of giardiasis have been associated with unusual manifestations, including pruritus, urticaria, uveitis, sensitisation towards food antigens and synovitis (Robertson et al. 2010). A study of manifestations of giardiasis in non-outbreak cases in the USA suggested that extra-intestinal symptoms are not particularly rare, being reported in over 30 % of participants enrolled in a study (Cantey et al. 2011).

Symptoms commonly start about 1 week after infection, and may continue until effective treatment, be self-resolving or become intermittent. It should be noted that intermittent symptoms contribute to delayed health-seeking behaviour and thus delayed diagnosis of infection (Cantey et al. 2011). Whether a particular symptom spectrum is more likely to be associated with Assemblage A infection or Assemblage B infection is unresolved, and geographical or population differences seem to occur (Robertson et al. 2010). Simultaneous or sequential infection with *Giardia* of different Assemblages has been proposed to result in more severe pathology, and thus more critical clinical presentation, than would be expected from a simple additive effect, with enhanced epithelial disruption and apoptosis (Koh et al. 2013). Additionally, size of the infectious dose may play a role (reviewed by Buret and Cotton 2011). Although the immune status of the host may contribute to the variable manifestations of giardiasis, the occurrence of refractory symptoms in otherwise healthy patients without giardiasis following treatment suggests that a more complex clinical picture probably occurs (Hanevik et al. 2007).

As giardiasis is predominant in children, some studies have investigated whether *Giardia* infection in children in developing countries results in long-term health consequences, particularly with respect to cognitive function and failure to thrive. Although the results of these studies have not always been completely concordant (Cacciò and Sprong 2011), there is accumulating evidence that in certain situations giardiasis in infancy or childhood may result in long-term disadvantages such as reduced cognitive function. For example, a study from Peru concluded that malnutrition in early childhood and potentially *Giardia* infection are associated with poor cognitive function at 9 years of age (Berkman et al. 2002). Furthermore, the authors concluded that should these associations be causal, then intervention programmes designed to prevent malnutrition and giardiasis in early life could lead to a significant improvement in the cognitive function of children in lower income communities throughout the less developed world (Berkman et al. 2002).

Infection with *Giardia* results in antibody production, particularly IgA that is secreted into the lumen of the intestine. However antigenic variation, specifically the ability of *Giardia* to alter the trophozoite surface proteins, and thus the epitopes to which the antibodies react (the variant-specific surface proteins or VSPs that are genetically determined) preclude or limit effective elimination of the parasite by antibody activity, although they may have an important role in limiting chronic infections. *Giardia* is the only organism inhabiting the intestine that has been demonstrated to have the ability to replace its surface proteins (Nash 2011), and in culture studies have demonstrated that this switching occurs spontaneously, in the apparent absence of environmental triggers, although if antibodies directed to the amino-terminus of the VSPs are cytotoxic, then the VSPs are replaced by others that are unrecognised by the host. However, several other immune defence mechanisms are also activated during *Giardia* infection and can contribute to the elimination of infection (Singer 2011); these include mast cells, complement, antimicrobial peptides such as defensins and nitric oxide. In addition, bacterial secretions and immune responses that have been activated by other infections may be effective in reducing the extent or the duration of infection (Singer 2011).

Diagnosis of giardiasis is usually based upon demonstration of cysts (and, less frequently, motile trophozoites—although unless recently voided stools are kept warm, motility will not be observed) in faecal samples, or sometimes in duodenal aspirates. Although serological (antibody-based) detection is also possible for diagnosing giardiasis, the utility of such an approach has been questioned (Smith and Mank 2011). This is because (a) different isolates have different antigenic identities; (b) in chronic disease, immunodepression may be encountered; and (c) antigenic variation down-regulates antibody production. While a common, nonvariant, immunodominant antigen suitable for serodiagnosis has yet to be identified, it has been suggested that *Giardia*-specific α-giardins might be of value for diagnosis of acute infections (Smith and Mank 2011).

For faecal samples, a concentration technique such as formol-ether (ethyl acetate) or flotation (often sucrose or zinc sulphate) is usually used prior to microscopy, which may be direct light microscopy, but may also involve staining. Stains that are used frequently include chlorazol black stain, Giemsa staining or, if a fluorescent microscope is available, immunfluorescent antibody test (IFAT), in which the parasite cysts are labelled with an antibody with a fluorescent tag. Antigen tests, such as ELISA-based assays, have also been developed, and rapid tests based on the same principle (immunochromatographic assays) are also commonly used. The disadvantage with such rapid assays is that not only are they relatively expensive, but also they may be of low sensitivity, particularly if cyst numbers are low (Strand et al. 2008). However, they are very simple to use and can provide a result within minutes at the point of care, and thus are popular amongst some users.

It is generally accepted that screening for intestinal parasites, including *Giardia*, by PCR will become more and more common in the next decades, and, as the feasibility improves due to automation and high-throughput facilities, might even replace microscopy of faecal concentrates (Stensvold et al. 2011). The specific detection of parasite DNA in stool samples using real-time PCR is particularly likely to become a method of choice, especially a multiplex approach allowing simultaneous testing for a range of different pathogens. However, microscopy of faecal concentrates currently remains a cornerstone, not only because many diagnostic laboratories do not have the technological capabilities for PCR, but also because of some limitations in extracting DNA from faecal material; when formalin has been used as a storage medium or for formol-ether sedimentation then this inhibits PCR, but also faecal samples themselves can include a range of PCR inhibitors including bilirubin, bile salts and complex polysaccharides (Smith and Mank 2011). Furthermore, because the specificity of primers and probes means that the only sequences detected will be those from which they were designed. Thus an unusual genotype may not be identified if the primers selected are too specific. Care should be taken in selecting primers and loci.

Thus, microscopy remains a mainstay of *Giardia* diagnostics, and it is important that, despite the progressive march of molecular technologies, such skills are maintained. The amount of information that can be obtained by microscopy should not be underestimated or undervalued, and, for a high cyst excretor, can result in a positive diagnosis being obtained within minutes.

It should be noted that analysis of more than one sample may be important, particularly for detecting mixed infections and determining their clinical importance (Smith and Mank 2011). This is especially so if infection and cyst excretion dynamics, symptoms and health effects vary for different infecting Assemblages or genotypes, and lack of identification of mixed infections may have a negative effect on our understanding of the epidemiology and clinical relevance of different *Giardia* infections (Smith and Mank 2011). In regions where giardiasis is endemic, mixed infections may be the norm and could be the consequence of multiple sporadic infections or waterborne or foodborne outbreaks (Smith et al. 1995).

Although giardiasis can be effectively treated with drugs, albeit with some dis-comforting side effects, for some patients treatment is ineffectual, and various che-motherapeutic regimes must be tried (Robertson et al. 2010). The most commonly used treatment for giardiasis is 5-nitroimidazoles. Metronidazole was the drug of choice for decades for treatment of giardiasis, but tinidazole is now often recom-mended as first-line treatment in many countries due to its efficacy being similar or superior, fewer side effects and better compliance (Escobedo et al. 2010). Albendazole, mebendazole and nitazoxanide are also commonly used, being well tolerated and with the added advantage of also being effective against intestinal helminths. In refractory cases, a treatment ladder is usually tried, and quinacrine (in combination with metronizadole) is often effective as a last resort, but may result in some unpleasant side effects (Mørch et al. 2008). Prolonged abdominal and fatigue symptoms have also been reported in patients, even after successful treatment in which the parasite itself is eliminated (Robertson et al. 2010). The reason for these prolonged symptoms is not fully understood, but may be due to changes in the architecture of the intestinal wall or alterations in the microbial flora.

Chapter 2
Transmission Routes and Factors That Lend Themselves to Foodborne Transmission

Transmission of *Giardia* infection occurs when an appropriate number of infectious *Giardia* cysts are ingested by a susceptible host. Transmission can be hand-to-mouth, and may be associated with unhygienic conditions or high-risk behaviour (Escobedo et al. 2010). Although sporadic cases of giardiasis in the community can be of individual clinical significance, the major public health importance of giardiasis lies in the potential for outbreaks to occur when drinking water, recreational water or food become contaminated with infectious *Giardia* cysts. Such contamination can result in several individuals becoming infected via the same transmission vehicle, and, for drinking water in particular, this can be of considerable community and economic importance, with, hundreds or even thousands of people at risk of infection (Robertson and Lim 2011). Additionally, when a large-scale outbreak occurs, with many infections occurring simultaneously in a particular community due to contamination of a common vehicle, the potential for subsequent environmental contamination increases accordingly, and thus the potential for secondary spread (Robertson et al. 2008).

Particular factors in the biology of *G. duodenalis* mean that this parasite is particularly suited to foodborne or waterborne transmission. These are the following:

- The large number of infective cysts that are excreted by an infected individual into the environment (numbers of between 1.5×10^5 and 2×10^6 cysts per gram of faeces have been quoted: Robertson and Lim 2011)
- The relatively low infectious dose
- The robustness of the cyst and its ability to survive in the environment; experimental results suggesting that cyst viability is retained for at least a month in damp conditions and in the absence of freeze–thaw cycles (DeRegnier et al. 1989; Robertson and Gjerde 2006), and that cysts are to some extent resistant to commonly used disinfectants such as chlorine
- The relatively small size of the cysts (8–12 µm in length) that enables penetration of sand filters used in the water industry

L.J. Robertson, *Giardia as a Foodborne Pathogen*, SpringerBriefs
in Food, Health, and Nutrition, DOI 10.1007/978-1-4614-7756-3_2,
© Lucy J. Robertson 2013

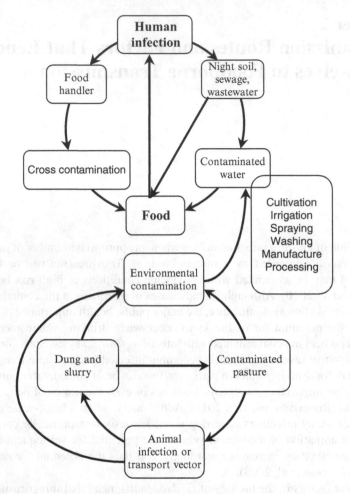

Fig. 2.1 Routes of contamination of food, particularly fresh produce, with *Giardia* cysts (adapted from Robertson and Chalmers 2013 for *Cryptosporidium*)

- The possibility for zoonotic transmission—this means that there is greater potential for environmental spread and contamination, and also for the amplification of cyst numbers by animals (including wildlife such as beavers) living in a catchment area contaminated by human faeces (Kutz et al. 2009)
- The possibility for onward contamination or transfer by transport hosts such as insects; promiscuous-landing synanthropic flies have been particularly associated with the carriage of protozoan parasites to food (Conn et al. 2007)

Taken together, not only do these factors mean that there is a high potential for possible vehicles of infection such as food or water with *Giardia* cysts, but also that they will probably survive on such vehicles in sufficient quantities to pose an infection risk to susceptible hosts. A visual representation of foodborne transmission possibilities, with emphasis on fresh produce, is provided in Fig. 2.1.

It is worth noting that in a recent risk ranking of foodborne parasites (http://www.fao.org/food/food-safety-quality/a-zindex/foodborne-parasites/en/ and http://www.who.int/foodsafety/micro/jemra/meetings/sep12/en/; see also Robertson et al. 2013), *Giardia* was ranked as number 11 out of 24 potentially foodborne parasites in terms of importance as a foodborne pathogen. This reflects not so much the relatively low number of cases of foodborne giardiasis that have been documented in the scientific literature, but the realisation of the potential for this parasite to be transmitted via the foodborne route. In this risk-ranking exercise, fresh produce is listed as the food commodity that is most likely to act as a transmission vehicle for giardiasis. However, it should be noted that when the documented outbreaks of foodborne giardiasis are considered, it is apparent that a wide range of foods have the potential to act as effective vehicles of transmission.

Chapter 3
Documented Foodborne Outbreaks of Giardiasis

Waterborne transmission of giardiasis is well known, and outbreaks of waterborne giardiasis have been extensively documented. Indeed, as waterborne transmission has the potential to result in infection of a larger number of people than foodborne transmission, and as *Giardia* cysts that contaminate water are probably more likely to remain infectious than cysts that contaminate food products (as *Giardia* cysts survive best under moist, cool conditions), most research interest and funding have been directed towards this transmission route. Waterborne outbreaks have been extensively reviewed (Karanis et al. 2007; Baldursson and Karanis 2011), with over 200 documented, the first reports being from the mid-1950s. Most outbreaks of waterborne giardiasis are reported from developed countries—indeed the largest outbreak of waterborne giardiasis in recent times is reported from Bergen, Norway (Robertson et al. 2006), a country considered to be one of the wealthiest in the world—however, common sense tells us that waterborne giardiasis is probably more likely to occur in less developed countries, where giardiasis is more likely to be endemic and where those infrastructures that are necessary for ensuring a safe drinking water supply, such as an intact sewage disposal system, effective catchment control measures and efficient water treatment, may be suboptimal (Robertson and Lim 2011). That outbreaks or cases of waterborne giardiasis are rarely reported from less developed countries is probably more related to the endemicity of giardiasis, which makes detection of outbreaks and vehicles of transmission difficult, and a lack of detection and monitoring systems, both at the public health level and at the water treatment level. A waterborne outbreak would probably be less likely to be detected in a community where a considerable proportion of the population are already infected with *Giardia*, because, unless the outbreak cases are particularly distinctive, the outbreak cases would be unlikely to show up against the background of non-outbreak cases.

That waterborne giardiasis is of greater public health significance than foodborne giardiasis is also reflected in the fact that standard methods for analysis of water for *Giardia* cysts were first developed between 5 and 15 years ago (e.g. the US EPA Method 1623; ISO Method 15553), whilst, as of today, there is no widely

L.J. Robertson, *Giardia as a Foodborne Pathogen*, SpringerBriefs
in Food, Health, and Nutrition, DOI 10.1007/978-1-4614-7756-3_3,
© Lucy J. Robertson 2013

Table 3.1 Documented outbreaks of foodborne giardiasis, including probable vehicle of infection (based on Robertson and Lim 2011)

Associated food matrix	Probable source of contamination	Estimated number of cases	Reference
Christmas pudding	Rodent faeces	3	Conroy (1960)
Home-canned salmon	Food-handler	29	Osterholm et al. (1981)
Noodle salad	Food-handler	13	Petersen et al. (1988)
Sandwiches	Unknown	88	White et al. (1989)
Fruit salad	Food-handler	10	Porter et al. (1990)
Tripe soup	Infected sheep?	–	Karabiber and Aktas (1991)
Ice	Food-handler	27	Quick et al. (1992)
Raw sliced vegetables	Food-handler	26	Mintz et al. (1993)
Oysters	Unknown	3	Smith-DeWaal et al. (2001)

accepted standard method for investigating food products for these parasites. Nevertheless, the value of monitoring of drinking water (either post- or pretreatment) for *Giardia* contamination has been the subject of considerable debate, as the methods are both expensive and time consuming, and interpretation of data can be difficult (for fuller discussion see Robertson and Lim 2011). However, whilst it is generally agreed that general performance indicators (e.g. turbidity, particle removal, pressure in distribution system) are probably of most importance for ensuring the microbial safety of the drinking water supply, it is also acknowledged that regulatory, event-driven monitoring of source water for contamination, using a site-specific monitoring programme, may provide important data that can be used as data input for risk assessments for an individual water source and thereby enable the application of appropriate barriers. Analysis of water samples for *Giardia* cysts may also supply critical information in the event of an outbreak.

Despite the potential for foodborne transmission of giardiasis, there is relatively little information available on this topic, and, to date, only nine outbreaks of foodborne giardiasis have been thoroughly documented, affecting approximately 200 people in total (see Table 3.1). In addition to the outbreaks listed in the table, which have all been described in the peer-reviewed scientific literature, a further 16 (affecting some 350 individuals) are listed on the Foodborne Outbreak Online Database run by the Center for Disease Control and Prevention, http://wwwn.cdc.gov/foodborneoutbreaks/Default.aspx, which lists outbreaks from 1998 onwards (the oysters outbreak listed in Table 3.1 is also included in this database), and for which the aetiology is confirmed for 15 outbreaks, but only suspected for one outbreak involving four individuals. Of these further 16 outbreaks, 6 are restaurant related, whilst 6 others are associated with various other types of community setting (office, religious, school, camp). For one of these outbreaks, one person is reported to have been hospitalised. This database also indicates that foodborne outbreaks of giardiasis are infrequently reported in the USA: during the period 2000–2010, of the foodborne outbreaks listed in the database, and with an identified aetiology, less than 1 % were attributed to *Giardia*.

For most of the 16 outbreaks included in the CDC database, the vehicle of infection is not identified, but unspecified vegetables, chicken salad, lettuce-based salad and multiple foods are listed for others. This database, which is continually updated, along with the information provided in Table 3.1, indicates that a whole range of different foods have the possibility to act as vehicles of transmission for giardiasis. As *Giardia* cysts are inactivated by desiccation and heat treatment, it is those food substances that are intrinsically moist and are most often consumed raw or very lightly cooked, such as salad vegetables, fruit, shellfish or dairy products, which would seem to be the most likely vehicles for infection. Whilst these food products have indeed been associated with infection transmission, and consumption of green salad, lettuce or other raw vegetables has been identified as a risk factor for acquiring *Giardia* infection in England, Germany, Cuba and Malaysia (Stuart et al. 2003; Espelage et al. 2010; Mohammed Mahdy et al. 2008; Bello et al. 2011), other less obvious foods are also listed as being associated with outbreaks of giardiasis, such as tripe soup and Christmas pudding. It should be noted that in those studies exploring risk factors for giardiasis, other variables have also been noted as risk factors for infection. For example, in the study from England (Stuart et al. 2003), in which a matched case–control study investigated 232 cases and 574 controls, other risk factors that were positively and independently associated with infection included (as well as consumption of lettuce) swallowing water when swimming, recreational contact with fresh water and drinking tap water. However, in the study from Germany, in which 120 laboratory-confirmed cases of autochthonous clinical giardiasis were compared with 240 randomly selected controls matched by county and age group (Espelage et al. 2010), cases were more likely to be male, immunocompromised and daily consumers of green salad—but, surprisingly, there was no association with exposure to water (swimming or water sports). In the study from Cuba (Bello et al. 2011), in which the focus was on hospitalized children, the only two independent, significantly associated variables of those investigated were eating unwashed vegetables raw and nail-biting, and the authors suggest that surveillance of drinking water and foodstuffs for *Giardia* and other parasites might help to reduce the hospitalisation of Cuban children. The study from Malaysia (Mohammed Mahdy et al. 2008) investigated *Giardia* infection among the Orang Asli people, and found that as well as eating raw vegetables being associated with infection, so was drinking piped water, and the authors suggest that both factors should be considered in future control strategies.

However, for some of these documented foodborne outbreaks it has been speculated that the vehicle of infection identified in the original report may be incorrect (Robertson 2013). One example is the outbreak recorded from 1991 in which a few people are reported to have been infected with *Giardia* as a result of eating soup made from sheep tripe (see Table 3.1; Karabiber and Aktas 1991). The authors of the outbreak report suggest that the source of infection could have been the sheep tripe and that the parasite could have been protected from inactivation during the cooking process by the crevices of the intestine. Although this possibility cannot be disproven, our current knowledge suggests that *Giardia* in sheep is most likely to be of Assemblage E, and thus non-infective to humans (e.g. see Robertson 2009). Additionally, if the soup was heated thoroughly or boiled, then it seems unlikely that the structure of the intestine would protect the cysts from inactivation. It has been speculated that, in this case, a food-handler might be a more probable common source of infection (Robertson 2013). Indeed, it should be noted that food-handlers are considered to represent a common route of contamination of food (Greig et al. 2007) and that any item that is handled by an infected food-handler with poor hygiene may act as an infection vehicle for giardiasis (Girotto et al. 2013). Various surveys of food-handlers for endoparasites in different countries have reported *Giardia* infection (e.g. Kamau et al. 2012 in Kenya; Siala et al. 2011 in Tunisia; Zaglool et al. 2011 in Saudi Arabia; Freites et al. 2009 in Venezuela; Babiker et al. 2009 in Sudan; Gündüz et al. 2008 in Turkey; Khurana et al. 2008 in India; Andargie et al. 2008 in Ethiopia; Costa-Cruz et al. 1995 in Brazil). Although at least two of these studies investigated fingernail scrapings for parasites, and found them nega-tive, the fact that food-handlers are infected and excreting cysts indicates the pos-sibility for contamination of food when there is a breakdown in hygiene.

An earlier outbreak of giardiasis, in which Christmas pudding is suggested as the vehicle of infection (see Table 3.1; Conroy 1960), can be less easily explained as being the result of inadequate hygiene associated with a food-handler, as impressive and extensive analyses were undertaken by the manufacturer of the pudding. In these investigations, in which sub-portions of centrifuged pudding sediment were stained with Lugol's iodine and examined by microscopy, large quantities of *Giardia*-like cysts were identified in the sediment and the assumption was made that the pudding had been contaminated with faeces from an animal, probably a mouse, harbouring *Giardia* infection. Presumably such contamination must have happened subsequent to any cooking procedures that would have rendered the cysts non-infective.

Whilst it may be relatively easy to implicate a food-handler for contaminating food, and a food-handler may indeed be the most likely source of contamination in some outbreaks, particularly those in which multiple foods are associated, and espe-cially under a buffet or canteen situation, contamination of vegetables and fruits can, of course, take place at any point along the field to fork continuum. Irrigation water and splash up from the soil are both particularly relevant potential contamina-tion sources for fresh produce, while shellfish, such as oysters, also have the poten-tial to be contaminated *in situ* before harvesting. As shellfish in particular, but also fresh produce, are usually kept cool and moist before consumption, *Giardia* cysts probably have similar chances of surviving on such products as they do in a water body. However the number of people likely to be exposed to a contaminated food product is probably considerably less. Additionally, when a solid food product is contaminated, then the contamination may be localized to a particular area or por-tion of the product such that not all consumers of the same product, even when derived from the same lot, are necessarily exposed. This potential for localization of contamination provides a further confounder for epidemiological investigation of potentially foodborne outbreaks.

It should be emphasized that not all countries have a system in place for report-ing foodborne diseases, and, even in countries where such a system is established, under-reporting is considerable, largely due to lack of knowledge of the physician or the victim regarding the possible aetiological role of foods, particularly for parasitic infections such as giardiasis (Robertson and Lim 2011). Additionally, once a suspected contaminated food product has been eaten or discarded, then it is unavailable for analysis and thus confirmation of the foodborne route of infection becomes impossible. Whilst for an outbreak epidemiological analysis may enable the investigator to pinpoint a particular food item, for individual cases or a cluster of a small number of cases, this may be impossible. Indeed, even in large outbreaks it may be difficult to determine a suspect food, particularly with a buffet-type situation involving multiple food types and different combinations.

Chapter 4
Approaches to Detecting *Giardia* Cysts in Different Food Matrices

While diagnosis of *Giardia* infection can be based upon detection of antigens in faecal samples (or, more rarely, detection of antibodies in blood) detection of contamination of vehicles of infection with *Giardia* cysts, whether water or food, relies on isolating and identifying either the *Giardia* cysts themselves or DNA from the *Giardia* cysts, on the contamination vehicle. Additionally, it should be noted that the concentration of *Giardia* cysts in a faecal sample from an infected individual is likely to be considerably higher than cysts on a contaminated potential vehicle of transmission, and thus detection may be much more difficult, and therefore diagnosis in faecal samples and detection in food samples are not entirely comparable.

As previously mentioned, standard methods for detecting *Giardia* cysts in drinking water have been available for many years. In brief, these methods involve concentrating particles that are approximately the size of *Giardia* cysts (or larger) from a relatively large water sample (minimum of 10 l) by filtration (flocculation and sedimentation can also be used, but are used less frequently), eluting these particles from the filter into a smaller volume, concentrating the smaller volume (often by centrifugation), and then isolating the cysts from other material in the concentrate before detection. While immunomagnetic separation (IMS) is most commonly used for isolation, other separation techniques may be used such as density gradient flotation. Detection is usually performed by drying the final concentrate of around 50 µl onto a microscope slide and examining it with IFAT. The fluorescent marker on the monoclonal antibody used is usually fluorescein isothiocyanate (FITC) and, additionally, a stain for the nuclei, usually 4′,6 diamidino-2-phenyl indole (DAPI) is usually used for improving detection, before examination by fluorescent microscopy. However, other detection methodologies are possible, including using molecular based detection systems.

In principle, the approach to analysing food matrices for contamination with *Giardia* cysts is the same as that used for water. However, apart from for beverages, filtration of a large volume is not possible, and instead some sort of elution procedure must be used for the food item itself. This is likely to mean that a relatively smaller (in terms of portion size) amount of product can be analysed, and the

L.J. Robertson, *Giardia as a Foodborne Pathogen*, SpringerBriefs
in Food, Health, and Nutrition, DOI 10.1007/978-1-4614-7756-3_4,
© Lucy J. Robertson 2013

approach to elution is likely to be influenced by the physical and biochemical nature of the product, in order to optimise removal of the parasites into a fluid phase, but at the same time minimise contamination with particulates or other material that may hamper the following steps in the procedure. The variations in both the biochemical and physical characteristics of different matrices, from shellfish to meat, to dairy products and to fruits and vegetables, means that a common "one-method-suits-all" approach is unsuitable, and a method that is appropriate for one type of food is likely to result in recovery efficiencies being suboptimal in other matrices.

One issue that is of importance for isolation of *Giardia* cysts from environmental matrices, including food and water, is that IMS is frequently used. The IMS system for *Giardia* cysts first appeared on the market in around 1995, approximately 1 year after the IMS system for *Cryptosporidium* oocysts. Whilst a number of manufacturers have succeeded in developing their own IMS systems for *Cryptosporidium* oocysts, for *Giardia* cysts one particular supplier dominates the market to the exclusion of all others. While this presumably is simply a reflection of the challenge in developing an IMS system for *Giardia* system that works adequately, it does mean that market forces are unable to prevail regarding price, and IMS remains a very costly step in analysis. Although some laboratories have managed to exclude IMS from methods in order to reduce costs, for the majority of laboratories IMS has been shown to increase recovery efficiencies, probably as a result of the selective concentration inherent in the technique. The development of a competitor IMS system by another manufacturer would be positive for laboratories undertaking such analyses, not only from an economic perspective but also for ensuring product quality and development.

The final step in analysis is detection. For all the food types listed below, and also for the standard methods for analysing water, IFAT usually remains the method of choice and is the method stipulated in the ISO Method currently under development for analysis of specific food matrices for *Giardia* (and *Cryptosporidium*). Given the advances in molecular detection systems within recent decades, it may be surprising that IFAT (which is basically a microscopy-based detection system) has not been supplanted by a technique such as qRT-PCR or LAMP, particularly as the equipment required for IFAT, a fluorescence microscope, is highly expensive whilst PCR equipment is becoming more competitively priced. In the diagnostic lab, multiplex qRT-PCR is more often becoming the method of choice for protozoa diagnostics, including *Giardia* and other parasitic pathogens in stool samples (Stark et al. 2011; Taniuchi et al. 2011). However, in diagnostics the number of a particular pathogen in a sample is expected to be relatively high, whereas in environmental samples not only are the numbers low, but it may also be important to detect non-nucleated parasites (obviously not detectable by methods that rely on detection of nuclear material). This is because although non-nucleated *Giardia* cysts are of no public health importance in themselves, they do indicate that the material being investigated has been contaminated with *Giardia*, and that another subsample may contain nucleated, viable parasites. Additionally, the range of potential inhibitors in different environmental samples, including food matrices, is less well known, and differences in matrix types may require that PCR conditions are adjusted per matrix. Nevertheless, some research groups are beginning to publish on occurrence of parasites, including *Giardia*, in food matrices in which PCR is used for the detection system. For example, in a survey of ready-to-eat packaged greens in Canada, Dixon et al. (2013) used nested-PCR as the detection step for different parasites, including *Giardia*, subsequent to a simple washing step and concentration by centrifugation (without the use of IMS for purification). Unfortunately, this research apparently did not include any seeding experiments to determine the limits of detection, and although all samples that were found to be positive by PCR were also examined by IFAT, the researchers apparently made no effort to determine whether some samples might have been positive by IFAT and not by PCR (by examining PCR-negative samples by IFAT). Another study compared the use of PCR, IFAT and flow cytometry for detecting *Giardia* cysts eluted from fresh produce (lettuce and water spinach leaves) and also in irrigation water in Thailand, and found that IFAT and flow cytometry provided similar results, but that PCR often failed (Keserue et al. 2012). Thus, although these research reports indicate the possibility for using molecular methods in such surveys, until more comprehensive research comparing detection methodologies is undertaken and successfully adopted by different laboratories, it is probable that IFAT will continue to be the detection method of choice for the immediate future.

4.1   Fruits and Vegetables

As both common sense and outbreak considerations suggest that fresh fruits and vegetables are food products that have a relatively high likelihood of being vehicles of infection, in method standardisation most focus has been placed on these (particularly, leafy vegetables and berry fruits) by the relevant ISO Group (ISO/TC 34/SC9/WG6), although other food matrices (specifically fruit juice, milk, molluscs and sprouted seeds) have also been considered. However, it was concluded that either the requirement for a standard method for analysing these other food matrices for *Giardia* cysts was insufficient at the time of consideration (2011) or the data available in the scientific literature were insufficient to use as a basis for a standard method development. Thus, leafy vegetables and berry fruits have been the sole focus for analytical method standardisation at present (registered in the ISO/TC34/SC9 work programme with the number ISO 18744). The purpose with method standardisation is to attempt to ensure that not only comparable methods are used but also the methods used, provided that they are conducted by competent and appropriately trained personnel, are likely to provide satisfactory, robust recovery efficiencies. One difficulty with the establishment of a standard method is that it may become more difficult to innovate and improve upon a method. Once a method has been set as a standard, then any alterations in the method have to undergo a range of independent trials to show equivalency, and, unless there is a good incentive for this, the effort and costs may not be worth it for many laboratories. Thus, ongoing review and evaluation of standard methods should be a part of the remit. Although this may be the intention, again the costs involved may mean that people with the qualifications and experience to review standard methods against innovations do not have the motivation to undertake this task.

The method being considered for standardised analysis for leafy vegetables and berry fruits for *Cryptosporidium* and *Giardia* (ISO Draft 18744) is based very closely on the standard water protocol (ISO 15553), although with elution into a specified medium as the first step, either through agitation of the produce by shaking or by stomaching using a stomacher. Elution from the food is followed by concentration of the eluate by centrifugation, isolation of the parasites from other debris by IMS and finally identification by IFAT and DAPI staining, as previously described. The method development was based not only on the water analysis protocol but also upon publications that described the use of this method, or variations of this method, with recovery efficiencies considered to be acceptable; in particular the work of Cook et al. (2007) should be noted, as this group was the first to recommend elution into glycine rather than a detergent-based elution solution.

Leafy vegetables and berry fruits have two different challenges for elution procedures. For leafy vegetables (e.g. lettuce), there is a large surface area that has the potential for contamination, and for some varieties of lettuce the leaves are deeply lobed and frilly (e.g. oak leaf varieties) such that some leaf areas are protected. For other lettuce varieties, such as Mâche (also known as lambs' lettuce), rosettes of leaves are held together in nubs of roots, providing pockets for contaminants to gather and not be readily eluted. Additionally, such pockets are also likely to include soil and other debris that may be an impediment in further steps of the analytical procedure.

For berry fruits, the area for contamination is not the problem, but the delicacy of the fruit means that vigorous elution is likely to break the fruit themselves, and the resultant fruit tissue fragments in the elution liquid are likely to impede the subsequent concentration and isolation of *Giardia* cysts. Some fruits also have hairy, rather than smooth, skins, and it may be less easy to remove *Giardia* cysts from such berries. Indeed, experiments comparing attachment of *Toxoplasma* oocysts to smooth-skinned blueberries and hairy raspberries demonstrated that they were more likely to remain attached to the raspberries (Kniel et al. 2002). Thus, not only are such fruits perhaps more likely to be contaminated at consumption (parasites not removed by standard washing), but also it will be more difficult to elute the parasites from such fruits for identification.

The composition of the elution solution that is used has also been the subject of debate. Earlier research used the same solution that was recommended by the US Environment Protection Agency for removing parasites from membrane filters during analysis of water. This solution contains a mixture of salts, together with Tween 80, sodium dodecyl sulphate and antifoam A (Robertson and Gjerde 2000, 2001a), but other publications have found that 1 M glycine at pH 5.5 acts as a satisfactory elution medium (Cook et al. 2007; Amorós et al. 2010). This may be simpler to make than a complex detergent solution, and therefore is advantageous; however there is debate regarding the importance of pH, and whether different pH are more efficacious for different product types. This issue is currently not resolved and it is unclear why pH should impact on removal efficiency for different types of produce.

The quantity of sample analysed is also a matter for consideration; for water samples the volume analysed is considerably over a portion size (minimum of 10 l), but for food samples, it is probably not possible to analyse in an equivalent fashion and it has been demonstrated that the greater the sample size, the less efficient the method at recovering parasites (Robertson and Gjerde 2001b). This reduced efficiency is presumably a reflection of both compromised elution efficiency and the increased quantity of other debris from the produce in the eluate that may inhibit or hinder other steps in the analysis process. Thus sample size should be a compromise that is selected to maximise recovery efficiency and also to enable detection of low-level contamination, at least at the infective dose per portion size. This is likely to vary according to the type of sample being analysed, but currently recommended sample sizes according to the ISO 18744 Method are between 25 and 100 g.

Although the standard method being developed is intended for analysis of berry fruits and leafy vegetables, analyses based on this approach have also been used for other varieties of fruits and vegetables including asparagus, baby sweet corn, Brussel sprouts, carrots, chillies, peppers, herbs (parsley, dill), leeks, mushrooms, onions, tomatoes, water spinach and sprouted seeds of different varieties (alfalfa sprouts, mung bean sprouts and radish sprouts). The latter of these is considered a particular challenge to analyse (Robertson and Gjerde 2001b), as material washed from the sprouted seeds (particularly mung beans) during the elution step not only appears to clog the mesh in homogeniser bags but may also interfere with the IMS procedure, possibly by coating the paramagnetic beads and thereby reducing the binding potential between the antibody and the cysts. Such materials may include mucopolysaccharides from sprout or seed cell walls, other cellular debris, microflora and excretory products of microflora and biofilms, in particular bacterial exopolysaccharides (Robertson and Gjerde 2001b). As such materials are likely to increase in older samples, the freshness of the sample (not just for sprouted seeds) is also an important factor that might affect recovery efficiency of the method. In the ISO method currently under development it is stated that the sample should be regarded as perishable and analysis shall commence no later than 24 h after receipt. In addition, the samples should be stored refrigerated (between 4 and 8 °C) in order to reduce sample deterioration.

4.2   Shellfish

Although only one small outbreak of giardiasis associated with bivalve molluscan shellfish has been documented (see Table 3.1; Smith-DeWaal et al. 2001), this product group is recognised as having potential as a vehicle for transmission. Not only are such shellfish traditionally consumed raw or lightly cooked, but they are also likely to come into contact with protozoan cysts in sewage outflow or run-off from land due to their preferred locations (intertidal or estuarine areas or areas close to the coast). Pathogens in such waters may become accumulated in bivalve molluscan shellfish tissues due to their particular method of alimentation that involves filtration of large volumes of water and concentration of particles (Robertson 2007; Gómez-Couso and Ares-Mazás 2012).

Although a widely accepted, optimised method for analysis of shellfish for *Giardia* contamination has yet to be described (Robertson 2007), some research groups have attempted to develop an optimised method by artificially contaminating shellfish and comparing recovery efficiencies of different methods and approaches of analysis. Different research groups have sometimes reported rather different efficacies of very similar methods. In general, the methods start with a tissue homogenisation step (although some research group have used gill-washing or haemolymph), followed by a concentration procedure (usually centrifugation); most researchers apparently agree that tissue homogenates provide better results than gill samples (homogenate or washings) or haemolymph (Robertson and Lim 2011). Following concentration, a purification/cyst isolation procedure is used, which may be non-specific (flotation on caesium chloride or sucrose gradients, or lipid extraction) or specific (IMS). Although IMS has been considered useful by the majority of researchers, others have found that its performance is so severely affected by the nature of the matrix that it provides no advantages (Schets et al. 2007). In these cases it would appear that the users are experiencing a similar problem to that reported for sprouted seeds, and that some constituent of the shellfish homogenate forms a coat on the paramagnetic capture beads, thereby either preventing or reducing the binding of the parasites to the antibodies, or hindering magnetic holding of the beads to the tube wall when held in the magnet. Oysters have been reported to be particularly difficult, and it is assumed that because oysters tend to be mucilaginous this may be the factor that inhibits IMS.

However, the very different biochemical nature of shellfish compared to water concentrates or washings from fruits and vegetables may indicate that a completely different elution approach may be more suitable. Based on the relatively high protein content of shellfish (8–20 % depending on shellfish species), Robertson and Gjerde (2008) developed a pepsin-digestion method which was based on the methodology usually used for the detection of *Trichinella* spp. larvae in meat or for recovering *Ostertagia ostertagi* larvae from the abomasal mucosa of cattle. This method was found to result in relatively high recovery efficiencies (70–80 %) when followed by IMS and detection by IFAT (Robertson and Gjerde 2008), and has since been modified by Willis et al. (2012), in which the protein digestion is followed by concentration by centrifugation and washing in detergent solution. This modified method has the advantage of being considerably cheaper and apparently results in very little loss in recovery efficiency. However, for some samples analysed by the modified method of Willis et al. (2012), a large pellet size precluded complete analysis, and this problem might perhaps have been resolved by the inclusion of a further purification step, not necessarily based upon IMS (e.g. flotation). Further comparative research with protein digestion may provide a method that may be considered suitable for standardisation when it has been validated in other laboratories.

Although for fresh produce (fruits and vegetables) most studies have relied almost exclusively on IFAT for detection, with shellfish some research groups have used other techniques or combined IFAT with other techniques such as fluorescent in situ hybridisation (FISH) and PCR. Although some research groups have found PCR to be less sensitive than IFAT, several authors consider that no one method is superior to another, but that the different techniques complement each other, and provide different types of information. It should probably be noted that the recovery efficiency of the same method may vary with species of shellfish, with particularly mucoid shellfish, such as oysters, being more likely to have lower recovery efficiencies.

4.3 Meat

Apart from the outbreak of giardiasis that was assumed to have been associated with tripe soup (see Table 3.1; Karabiber and Aktas 1991), no outbreaks or individual cases are recorded that are associated with ingestion of contaminated meat or meat products. Therefore, there has been very little research directed towards detecting *Giardia* cyst contamination of meat. Although, a study from India investigated goat meat samples for contamination with *Giardia* cysts using IFAT, Lugol's stain and PCR for detection, the sample preparation is difficult to follow, and no recovery efficiencies of the method are provided (Rai et al. 2008). In another unpublished study, *Giardia* has been reported to have been detected in various raw meats (chicken breasts, minced beef, pork chops) from retail outlets using PCR and IFAT for detection (Dixon 2009). However, the actual process used for detection is not supplied, and, again, nor is the recovery efficiency. It might be expected that, on the whole, methods and recovery efficiency results would be similar to those for *Cryptosporidium* on meat products (which has been the subject of more research studies), and the reader is referred to the companion Springer brief ("*Cryptosporidium* as a Foodborne Pathogen") for further information on approaches to analysing meat for protozoa as surface contaminants.

4.4 Beverages

Apart from outbreaks of giardiasis associated with drinking water, there have been no documented outbreaks, or individual cases, associated with beverages. Thus research directed towards detecting *Giardia* cysts as contaminants of beverages such as fruit juices, milk or other beverages has been relatively scarce. A study from India analysed 20 ml milk samples for *Giardia* cysts by using a method in which the sample was treated with 1 ml Bacto-Trypsin and 5 ml Triton X-100 for 30 min at 50 °C, before centrifugation and washing in water (Rai et al. 2008). However, no recovery efficiencies of the method are provided. Similarly, a study from Egypt on fruit juices in which analysis involved centrifugation, Sheather's sugar flotation and staining with modified Ziehl–Neelsen also gave no indication of recovery efficiency (Mossallam 2010). More extensive work on protozoa in beverages has been directed towards *Cryptosporidium*, and it is likely that methods similar to those used for

Cryptosporidium could be adapted for analysis of samples for *Giardia*. Therefore, the reader is referred to the companion Springer brief ("*Cryptosporidium* as a Foodborne Pathogen") for further information.

4.5 Water Used in the Food Industry

The food industry is a large water user; among other things, water is used as an ingredient, as an initial and intermediate cleaning medium, for conditioning raw materials (soaking, cleaning, blanching and chilling) and as a conveyor of raw materials. Prior to food-processing, water is used for irrigation of crops, for application of chemicals such as pesticides, for depuration and for general cleaning purposes. Although water of non-potable quality may be appropriate for some uses in the water industry, for other uses it is essential that the quality of the water is of the same microbiological quality as drinking water. In order to reduce water usage/wastage, in the fresh produce industry in particular, water might be recycled, often with a purification step (use of a sanitizer) to inactivate pathogens already removed from the produce (Gil et al. 2009). This sanitisation step often involves chlorination, but other technologies such as photocatalysis (Selma et al. 2008a), ozone and UV (Selma et al. 2008b) have also been proposed. The problem is that the efficacy of the sanitizing step is usually, understandably, directed towards bacteria, which have the potential to multiply on the produce. However, sanitizers that are effective against bacteria may be ineffective against protozoan cysts; although *Giardia* will not replicate in wash water, by reusing such water for a further washing step, cysts can be distributed onto previously clean areas of product such that a point contamination becomes spread throughout a batch, and thus a limited contamination that might be associated with the likelihood of a single case of infection may be distributed such that the possibility of a single case expands to the possibility of an outbreak.

Methods specifically directed towards analysing water used in the food industry for *Giardia* cysts have not been developed, but approaches based on the standard protocols for drinking water would probably be most appropriate (i.e. the US EPA Method 1623; ISO Method 15552), and have indeed been used for both irrigation water and processing water in the food industry (Robertson and Gjerde 2001a). It should, however, be realised that reused water may have a greater load of contaminating debris than drinking water, and thus a lower recovery efficiency may be expected. The use of an internal process control may be of use in such instances to monitor recovery efficiencies (Warnecke et al. 2003).

Chapter 5
Occurrence of *Giardia* Cysts in Different Food Matrices: Results of Surveys

The results of surveys for *Giardia* in different food matrices provide a snapshot in time of what was found on a particular food sample, under particular conditions, using a particular method, in a particular laboratory, by a particular analyst. As some of the methods are relatively expensive (particularly IMS, if used) often only a small number of samples are analysed, and as recovery efficiencies are often superior with smaller sample sizes, only a small quantity of the product is analysed. Thus, results from such surveys can only be considered to give a very diffuse insight into the risk of ingestion of a *Giardia* cyst from a particular product. Furthermore, surveys that do not investigate genotype of any *Giardia* cysts detected cannot even determine whether cysts that are found are infectious to humans, and thus of public health significance. Additionally, surveys that do not consider the viability or infection potential of the cysts are similarly hampered. Nevertheless, whilst acknowledging the limitations of such surveys, it should also be realised that these results are our only verified, scientific handle on contamination of food matrices with potentially infective *Giardia* cysts, and thus the information that they provide is useful. In addition, investigation of food matrices for *Giardia* contamination in an outbreak situation has the potential to identify infection routes, and thus take measures against them. Information from food products analysed under an outbreak situation is of greater value if survey data are also already available against which the outbreak-related analyses can be compared. Occurrence data are being improved all the time, as further studies are conducted with better, more efficient methods. The information provided in the sections below is intended to give an insight into what we know about the occurrence of *Giardia* cysts on different product types as of today—more up-to-date information should always be sought. When comparing occurrence results from different studies, the method used and the recovery efficiency of that method should always be borne in mind, as surveys using very different methods or similar methods but with different recovery efficiencies cannot properly be compared. When recovery data are not provided, it may be most appropriate to assume that it is low, and thus any contamination data provided are likely to be conservative. It should be noted that it is not appropriate to use recovery

L.J. Robertson, *Giardia as a Foodborne Pathogen*, SpringerBriefs
in Food, Health, and Nutrition, DOI 10.1007/978-1-4614-7756-3_5,
© Lucy J. Robertson 2013

efficiency data obtained by another research group, even if the analytical methods used are very similar. Examination of the literature reveals that different laboratories may achieve very different recovery efficiencies; for example, while Cook et al. (2007) report a recovery efficiency of *Giardia* cysts from artificially contaminated salad leaves of 46.0 ± 19.0 % ($n=30$), and of 36.5 ± 14.3 % ($n=20$) when internal controls were used on samples, a study using the same technique in Spain (Amorós et al. 2010) reported a mean recovery efficiency of *Giardia* cysts from salad vegetables (Chinese cabbage and lettuces) of only 16.7 ± 8.1 % ($n=8$)—this is less than 50 % than that achieved by the lab developing the method.

5.1 Fruits and Vegetables

Outbreaks and method development are probably the two greatest drivers for surveys for contamination of food products with *Giardia* cysts, and fresh fruits and vegetables have probably been the food matrices with the greatest number of surveys (Robertson 2013). Despite produce-associated outbreaks being relatively rare, *Giardia* cysts have been detected as contaminants on/in a range of raw vegetables and fruits (see Table 5.1). In general, a widespread, low-level contamination of fresh produce can be inferred from these results. In addition to the studies listed in Table 5.1, in which the analytical methodologies used are based upon the principles adopted by a proposed ISO standard, studies have also been conducted in nearly all regions of the world, including countries in Africa, Asia and South America where giardiasis is perhaps considered endemic; these studies (e.g. Amahmid et al. 1999; Vuong et al. 2007; Monge and Arias 1996; Takayanagui et al. 2000; Fallaha et al. 2012; Keserue et al. 2012) also tend to show widespread contamination in different fresh produce, and associations are occasionally made with crop cultivation variables (e.g. use of wastewater for irrigation or seasonality). For example, a survey of vegetables from supermarkets and a public market in the Philippines reported that 1 % of vegetables were contaminated with *Giardia* cysts, and it was speculated that application of human and animal waste to agricultural land was one possible contamination source (De Leon et al. 1992).

Interestingly, North America seems to be one region of the world where surveys of vegetables for *Giardia* cysts are lacking. However, a recently published survey from Canada (Dixon et al. 2013) goes some way to correcting this omission and reports a 1.8 % prevalence of *Giardia* (10 of 544 samples found positive) in packaged ready-to-eat leafy greens that included iceberg lettuce, romaine lettuce, baby lettuces, leaf lettuce, radicchio, endive and escarole, and with some samples also containing spinach and romaine lettuce. Most samples were grown in the USA, but some were from Canada and/or Mexico. However, this study used PCR for detection (with IFAT used to screen those samples that were PCR positive), but unfortunately did not provide any data on limits of detection. In addition, although two of the PCR-positive samples were found also positive by IFAT, the number of cysts detected was not documented (the authors state that in "the majority of surveillance studies" no

Table 5.1 Occurrence of *Giardia* cysts on fresh produce

Location of survey and origin of produce analysed	Type of fresh produce analysed	Results: proportion of samples contaminated with *Giardia* cysts	Concentrations of cysts on positive samples	Reference
Norway; produce both imported and locally produced (beansprouts grown locally from imported seed)	475 samples comprising alfalfa sprouts, dill, lettuce, mung beansprouts, mushrooms, parsley, precut salad mix, radish sprouts, raspberries, strawberries	Dill: 2/7 (29 %) Lettuce: 2/125 (2 %) Mung bean sprouts: 3/149 (2 %) Radish sprouts: 1/6 (17 %) Strawberries: 2/62 (3 %) *Giardia* cysts not detected in other samples	No. cysts per 100 g produce sample: Dill: 4–8 Lettuce: 2–3 Mung bean sprouts: 2–6 Radish sprouts: 2 Strawberries: 1	Robertson and Gjerde (2001a)
Palermo, Sicily	20 raw vegetable mixes (leafy vegetables and carrots), 20 leafy vegetable mixes	1 sample containing leafy vegetables and carrots positive for *Giardia*. All other samples negative	12 cysts per 50 g produce sample	Di Benedetto et al. (2007)
York, UK	20 raw vegetable mixes including various lettuce varieties, carrots, mange tout, spring onions, parsley, chillies, baby sweet corn, asparagus	1 sample containing organic water cress, spinach and rocket salad positive for *Giardia*. All other samples negative	1 cyst per 50 g produce sample	Cook et al. (2007)
Valencia, Spain	19 samples comprising Chinese cabbage, lollo rosso lettuce, romaine lettuce	Chinese cabbage: 2/6 (33 %) Lollo rosso lettuce: 3/4 (75 %) Romaine lettuce: 5/9 (56 %)	No. cysts per 50 g produce: Chinese cabbage: 1–9 Lollo rosso lettuce: 1–4 Romaine lettuce: 1–8	Amorós et al. (2010)
Norway; produce both imported and locally produced	41 samples comprising salad leaves (of different varieties), raspberries, and mangetout	Mangetout: 1/10 (10 %) Other samples negative	1 cyst per 50 g mangetout	Johannessen et al. (2013)

Studies included in table use methods based on principles adopted by a proposed ISO standard attempt is made to enumerate the parasites—although this is manifestly not the case for those studies using methods based on the principles adopted by the ISO Method; see Table 5.1). Providing a measure of the concentration of cysts detected per product is useful input data for risk assessment, and can be obtained with very litte extra effort when IFAT is used as the detection method; regular PCR (as opposed to real time-PCR) as used in this study provides only presence or absence information and no information on quantification. However, PCR analysis can provide useful genotype information that cannot be obtained by IFAT unless secondary, post-detection molecular analyses are performed; in the Canadian study, which Assemblage of *Giardia* was detected is described for nine of the positive samples, being Assemblage B for seven samples and Assemblage A for two samples. It is unclear why this information could not be obtained for one sample. These data suggest that not only are the cysts, if infectious, of health significance for the human consumer but also the sources of contamination are likely to be human rather than animal. Thus, as human sewage is unlikely to be used for fertilizer of food crops cultivated in the USA, it is tempting to suggest that the most likely sources of contamination here were either human handlers or water used for washing the produce in the packing plant.

There are also a considerable number of studies published in the "grey literature"—particularly from Asia, but these studies are often difficult to access and the quality of the study is sometimes questionable. For these reasons, these studies are not detailed or referenced here. Whether the cysts detected are infectious/viable and of a genotype of public health significance has not been explored in any study, presumably because the cysts tend to be found at low concentrations (this is the reason stated for not exploring viability in the study by Dixon et al. (2013)) and also, for many studies, the detection method inactivates them or reduces their viability (e.g. fixing to microscope slides).

5.2   Shellfish

Although some studies have suggested that *Giardia* cysts may not be ingested by shellfish as readily as *Cryptosporidium* oocysts (Graczyk et al. 2003), and also that if they are ingested they may be digested by the shellfish themselves (Graczyk et al. 2006), experimental studies have shown that some types of shellfish, such as clams and oysters, can concentrate *Giardia* cysts in their tissues (Gómez-Couso and Ares-Mazás 2012). Additionally, the association of *Giardia* cysts with marine macroaggregates has been speculated to enhance their bioavailability to invertebrates and thus their subsequent incorporation into the marine food web (Shapiro et al. 2013).

Although only three of the five surveys of shellfish (of species that are commonly eaten) for *Giardia* cysts published between 1997 and 2007 reported detection of *Giardia* cysts (Robertson 2007), a further five survey studies published since that review was published (Lévesque et al. 2006, 2010; Robertson and Gjerde 2008; Lucy et al. 2008; Leal Diego et al. 2013) have also all reported the occurrence of *Giardia* cysts (see Table 5.2). Additionally, some of the studies have investigated the genotype of the *Giardia* cysts detected, and established that, genotypically, they

Table 5.2 Occurrence of *Giardia* cysts in shellfish

Location of survey	Shellfish analysed	Results: proportion of samples contaminated with *Giardia* cysts	Cyst concentrations in positive samples	Reference
Chesapeake Bay, USA	*Crassostrea virginica* (Eastern oyster)	0/360 (0 %)	–	Fayer et al. (1998)
Commercially available in markets in Alexandria, Egypt	Gandofli (*Caelatura pruneri*), wedgeshell clam (Om el Kholool) (*Donax trunculus limiacus*)	0/180 (0 %)	–	Negm (2003)
Specimens originated from Spain, UK, Italy, Ireland, New Zealand	Mediterranean mussel (*Mytilus galloprovincialis*); common cockle (*Cerastoderma edule*); Manila clam (*Ruditapes philippinarum*); Pullet carpet shell (*Venerupis pullastra*); Smooth Artemis (*Dosinia exoleta*); European flat oyster (*Ostrea edulis*); banded carpet shell clam (*Venerupis rhomboideus*); warty venus shell (*Venus verrucosa*); green lipped mussel (*Perna canaliculus*)	Mussel samples: 0/42 (0 %) Clam samples: 0/18 (0 %) Cockle samples: 0/18 (0 %) Oyster samples: 1/9 (11 %)	PCR detection only—no enumeration	Gómez-Couso et al. (2004)
Four estuaries, Galician coast, Northwest Spain	Mediterranean mussel (*M. galloprovincialis*)	Pooled homogenate 77/184 (41.8 %)	1–19 cysts per sample	Gómez-Couso et al. (2005a)
Galician coast, Northwest Spain	Mediterranean mussel (*M. galloprovincialis*)	Pooled homogenate 83/200 (41.5 %)	Mean number of 58.3 cysts/sample	Gómez-Couso et al. (2005b)
Oosterschelde, Netherlands	Pacific cupped oyster (*Crassostrea gigas*)	6/179 (3.4 %)	Data not provided	Schets et al. (2007)
St. Lawrence River, Québec, Canada	Softshell clam (*Mya arenaria*)	26/41 (63.4 %)	Data not provided	Lévesque et al. (2006)
13 different sites on the Norwegian coast	Blue mussels (*Mytilus edulis*), European flat oysters (*Ostrea edulis*), and horse mussels (*Modiolus modiolus*)	Blue mussels: 7/14 (50 %) batches Horse mussels: 1/1 (100 %) batches Oysters: 0/1 (0 %) batches	Blue mussels: <1–4.5 per gram Horse mussels: <1 per gram	Robertson and Gjerde (2008)
Sligo Bay, Ireland	Blue mussels (*M. edulis*)	Homogenate from 105 mussels positive for *Giardia*	6.1 cysts per gram	Lucy et al. (2008)
Nunavik, Quebec	Blue mussels (*Mytilus edulis*)	2/11 samples (18.2 %)	Data not provided	Lévesque et al. (2010)
"Vale do Ribeira", Cananéia city, Brazil	Oysters (*Crassostrea brasiliana*)—both at collection and following depuration	At collection: 4/11 (36.3 %) After depuration: 6/11 (54.5 %)	Data not provided, PCR detection only for some samples	Leal Diego et al. (2013)

have the potential to infect humans (Gómez-Couso et al. 2004; Gómez-Couso and Ares-Mazás 2012), with *Giardia* cysts from both Assemblage A and Assemblage B reported. Viability/infectivity of the *Giardia* cysts detected has not been investigated in any studies. The variation in analytical techniques used for shellfish means that comparison between studies is extremely challenging, and it is impossible to reach any general conclusions, apart from that *Giardia* cysts appear to accumulate within various species of shellfish, and may have infectious potential. Thus, they cannot be excluded as a potential risk food with respect to transmission of giardiasis.

5.3   Meat

Currently there are few reports on investigation of meat for *Giardia* cyst contamination. One investigation that has been published is from India, where three batches of goat meat were considered to be positive by both IFAT (using a Cy3-labelled antibody) and PCR (Rai et al. 2008). Cyst enumeration does not seem to have been conducted and the photomicrograph provided in the publication is less than convincing. Although PCR detection was conducted, genotype data are not provided. The authors do not explore the likely source of contamination, and as no genotyping data are provided, it is difficult to speculate on whether contamination of goat meat is likely to be from faecal matter from the goats themselves, or from human handlers of the meat. Presumably contamination of meat via flies or other vectors is also possible.

In another unpublished study, *Giardia* was detected in various raw meats (chicken breasts, minced beef, pork chops) from retail outlets using PCR for detection, and one pork chop was also found to be positive for *Giardia* by IFAT (reported in Dixon 2009). One interesting finding from the Canadian data (Dixon 2009) is that genotyping of the positive samples indicated that the majority of contamination was from Assemblage B, which probably indicates a human source of contamination (food-handler)—although animal infections with Assemblage B have been occasionally reported (e.g. Lalle et al. 2005), it is much more common in human infections. However, in one minced beef isolate, the *Giardia* detected was of Assemblage E. This indicates an animal source of contamination (cattle, sheep), and it can be speculated that the most likely source of contamination in this case is from faecal/intestinal matter at the slaughterhouse. As Assemblage E is non-zoonotic, this should be considered to be of little public health significance. However, it does indicate that the product has probably been contaminated with animal faecal matter, and thus the potential for other animal pathogens—not just parasites—also being present.

5.4   Beverages

As with meat, there are few reports on investigation of beverages for *Giardia* cyst contamination. One study from India reported that two out of the three batches of milk were considered to be positive by both IFAT and PCR (Rai et al. 2008). But, as for the meat samples in the same study, information is not provided regarding the cyst concentrations or the genotypes of the cysts detected, and no information is provided that may indicate whether the cysts are more likely to originate from the animal being milked (it is not stated whether the milk is from cows or goats) or from human handling of the milk. This information is important both epidemiologically and regarding the likely infection risk to humans.

A study from Egypt investigated a range of different juices (strawberry, sugar cane, mango, lemon and orange) for contamination with protozoan parasites, and reports the detection of *Giardia* cysts in all juice types (Mossallam 2010). For each fruit juice, 35 samples were collected from roadside stalls for analysis and the occurrence of *Giardia* contamination ranged from 23 % (8 samples positive) for lemon juice to 9 % (3 samples positive) for orange juice. In total, 28 samples of juice (16 %) were reported to contain *Giardia* cysts, but cyst numbers per volume of sample analysed are not provided; such information would be relevant and useful for risk assessment. It is unclear why the number of *Giardia* cysts detected per sample is not provided. This study also attempted to assess cyst viability and infectivity, both by examining inclusion and exclusion of the fluorogenic dyes propidium iodide and fluorescein diacetate, and by mouse infectivity trials. The results from these studies indicated low cyst viability and lack of infectivity in the most acidic juices (orange (pH 2.9) and lemon (pH 3.2)), but high viability and infectivity in the other three juices that were less acid (ranging from pH 4 (mango) to pH 7.5 (sugar cane)). Although pH has been considered to be of relevance with regard to inactivation of *Giardia* cysts using disinfectants (Fernando 2009), and some acidic solutions have also been considered for use as "home" disinfectants (Sadjjadi et al. 2006; Costa et al. 2009), there are few precisely controlled studies on the effect of pH alone on the survival of *Giardia* cysts. Obviously, *Giardia* cysts survive the acidic pH in the stomach before excysting in the duodenum, but it is unclear if cysts exposed to a low pH continue to have infectious potential for a prolonged period.

5.5   Water Used in the Food Industry

Irrigation with untreated water is a major potential route of crop contamination with *Giardia* (Cook and Lim 2012). Surface waters may contain *Giardia* cysts, either from sewage discharge or from contamination from animal sources, and if this water is used for irrigation or for other agricultural uses (e.g. application of pesticides or fertilizers), then these may be transferred onto the surfaces of the crops. In Norway, samples from a river used for irrigation of lettuces were found to contain *Giardia* cysts (Robertson and Gjerde 2001a), although *Giardia* cysts were not detected on lettuces from this location that were analysed. A survey of irrigation waters at 3 sites where fruit and vegetable crops were produced in the USA, and at 22 sites in three Central American countries, demonstrated *Giardia* contamination at 2 (67 %) and 13 (59 %) sites, respectively, thus an overall prevalence of 60 % (Thurston-Enriquez et al. 2002). While the concentrations of *Giardia* cysts detected in waters in Norway were low (1 cyst per 10 l) and also relatively low in the USA (mean of 25 cysts per 100 l), in Central America the concentrations of cysts were considerably higher, with a mean of over 550 cysts per 100 l, indicating not only considerable contamination of the irrigation water but also, more importantly, a considerable potential for contamination of crops (Robertson and Gjerde 2001a; Thurston-Enriquez et al. 2002). In Mexico also the levels of *Giardia* cyst contamination in irrigation waters have been shown to be high, with over 50 % of the samples of irrigation water containing *Giardia* cysts, and with concentrations ranging from under 20 cysts per 100 l to over 1,600 cysts per 100 l (Chaidez et al. 2005), while in another study from a major irrigation system in Mexico in which six irrigation water samples were analysed for *Giardia* cysts, the mean concentration was as high as 3.5 cysts per 100 ml despite there being no direct sewage discharges into the irrigation canal system, indicating that contamination was probably from animals, weather events or localised contamination events (Gortáres-Moroyoqui et al. 2011). In developing countries, or countries where water resources are scarce, it makes sense to use wastewater for crop irrigation, and it has been shown that the use of untreated wastewater in agriculture can have major financial and nutritional benefits for farmers and consumers (Ensink and van der Hoek 2009). However, the potential for transfer of pathogens, including *Giardia* cysts, from the irrigation water to the crop should not be overlooked, and a simple pretreatment step may make a large difference to the contamination potential. For example, field trials from Morocco demonstrated that while *Giardia* cyst contamination could not be detected on different crops that were irrigated with wastewater that had been treated via waste stabilisation ponds, crops that had been irrigated with untreated wastewater were frequently contaminated with *Giardia* cysts, with approximately 40 % of coriander samples contaminated (mean cyst burden of 250 cysts per kg), 30 % of carrot samples (mean cyst burden of 150 cysts per kg), 50 % of mint samples (mean cyst burden of 100 cysts per kg) and 80 % of radish samples (mean cyst burden of 50 cysts per kg) (Amahmid et al. 1999). A study from Thailand has also suggested that flow-through canals, which can be viewed as waste stabilization ponds, are effective at removing *Giardia* cysts, with the main removal mechanisms considered to be sedimentation and sunlight (UV) irradiation (Diallo et al. 2009).

It should also be remembered that using untreated wastewater in agriculture may carry a risk to the farmers themselves as they are going to be in closer contact with a medium that is likely to contain pathogens. A study in Faisalabad, Pakistan, found that farmers that used untreated wastewater for irrigation were more likely to suffer from symptomatic giardiasis than farmers who did not (Ensink et al. 2006).

Cabbage and lettuce crops irrigated with sewage near Asmara, Eritrea, were also found to be contaminated with *Giardia* cysts (Srikanth and Naik 2004), with 50 % of samples positive. Similarly, water spinach grown near to wastewater discharge outlets near Phnom Penh, Cambodia, has been found to have relatively high contamination—*Giardia* cysts were detected on 56 % of 35 samples at a concentration of 6.6 cysts per gram (Vuong et al. 2007). A publication from a study in Thailand (Keserue et al. 2012) suggests that the researchers were able to "track" the cysts from wastewater to irrigation waters and finally to confirm the contamination of salads and water vegetables. In this study, concentrations of *Giardia* cysts in irrigation waters were around 10 cysts per litre, and a brief wash of the fresh produce by farmers prior to sale was considered to have only a very minor impact on removing the contamination. Note that unhygienic post-harvest handling (such as washing in contaminated water) might increase, rather than decrease, contamination levels (Ensink et al. 2007).

The method of irrigation with potentially contaminated water may facilitate or impede potential contamination of produce. For example, sprinkler irrigation, in which the water is piped to one or more central locations within the field and distributed by overhead high-pressure sprinklers or guns, is likely to result in contamination of the leaf-crop or fruit produce, whereas drip irrigation or trickle irrigation, in which water is delivered, drop by drop, at or near the root zone of plants, is less likely to result in contamination of leaves and fruits. Additionally, this method is probably more water efficient, as evaporation and run-off are minimised. Sub-irrigation, in which the water table is artificially raised so that the soil is moistened from below the plants' root zone, is also unlikely to result in contamination of fruits or leaves above the ground. Thus, ensuring that irrigation water is delivered to the roots, rather than coming into contact with the leaves and fruits of the plant, is likely to minimise contamination. Whilst this principle may be appropriate both for irrigation and application of fertilisers, it is probably unsuitable for application of pesticides. Additionally, for crops growing close to the ground (such as lettuce or strawberries), the potential for splash-up from contaminated soil may also be important. This may occur during irrigation, but should also be considered during intense rain. The potential for contamination of crops during, for example, flooding of fields during extreme weather events is one aspect of climate change that is beginning to be explored in research projects and also as a result of specific events (e.g. Casteel et al. 2006). However, the effect of extreme weather events on contamination of the drinking water supply with pathogens, including *Giardia*, is presently of greater focus (Kistemann et al. 2002; Cann et al. 2013) than contamination of produce during or following extreme weather events.

Also in the fresh produce industry, *Giardia* cyst contamination was detected in water used in bean sprout production in Norway (Robertson and Gjerde 2001a) (contamination assumed to have originated from the bean sprout seeds), whilst in Mexico 83 % of wash-water tank samples used in a packinghouse were found to be contaminated with *Giardia* cysts, with concentrations reaching as high as over 500 cysts per 100 l (Chaidez et al. 2005). Thus, spread of *Giardia* contamination between produce in produce washing facilities, particularly when wash-water is recycled between batches, is one potential aspect for contamination that should be considered as part of Hazard Analysis and Critical Control Point (HACCP) routines.

Depuration processes are another means by which water may be used to spread contamination from contaminated food, here shellfish, to non-contaminated food. During depuration, harvested shellfish are placed in a controlled aquatic environment, where it is intended that they will purge themselves of their gastrointestinal contents, and thereby any pathogens. The literature investigating whether depuration may result in spread of contamination with *Giardia* cysts, rather than reduction, is very scant and there are insufficient controlled studies to reach a conclusion

(Robertson 2007). However, the data that has been amassed suggest that depuration times (for *Giardia* cysts) vary between shellfish species, and standard times are probably ineffective (Nappier et al. 2010). In addition, other factors, such as temperature and salinity, may affect depuration and uptake of *Giardia* cysts in depuration tanks (Willis et al. 2013).

Processing procedures post harvest of crops may also involve contact with water, particularly washing procedures. Again, via such processing procedures, *Giardia* cysts may be spread from contaminated produce to non-contaminated produce, or throughout a batch with point contamination, or, if water is used that is already contaminated with *Giardia* cysts, then the contamination may be introduced. The extent to which such contaminations are spread, or become a potential risk to public health, depends upon a variety of factors including the concentration of cysts in the water, cyst viability, cyst infectivity as well as the specific contact and use of the water. In the fruit juices that were reported to be contaminated in Egypt (Mossallam 2010), the author speculates that the source of the juice contamination may be the water that was added to the juices rather than the fruit used for making the juice, as the juice that was subject to the least dilution with water (orange juice) had the lowest number of positive samples. However water that was used for dilution of the juice was not investigated for contamination and insufficient data are provided to reach a definitive conclusion. Again, other potential sources of contamination in this case could be the utensils or the containers that came into contact with the juice, or the handlers themselves. Although not directly concerned with *Giardia*, a study from Pakistan has indicated that unhygienic post-harvest handling was the major source of produce contamination for vegetables that had been irrigated with untreated domestic wastewater, and that an intervention at the market level, such as the provision of clean water for washing produce, could be a better way to protect public health and more cost effective than wastewater treatment (Ensink et al. 2007). It should be noted that a study from Iran found *Giardia* cysts on unwashed vegetables, but not on washed vegetables from the same villages (Shahnazi and Jafari-Sabet 2010), although other studies have noted that cursory washing of vegetables might have little impact on contamination with parasites (Keserue et al. 2012), and might even result in contamination in some instances.

Further research on the potential for water to act as a source of contamination of food is lacking. It might be noted that in major waterborne outbreaks of giardiasis, contamination of food by use of the contaminated water in the food industry has not, to date, been proven to be problematic. However, it is possible that infections may not be traced against a background of elevated infection. Also, in industrialised countries at least, food industries may have their own barriers in place (for example, in-line UV disinfection) to ensure that potentially contaminated water does not come into contact with vulnerable processes within the industry should municipal treatment fail or be inadequate.

Chapter 6
Inactivation or Decontamination Procedures

Giardia cysts are known to be robust. They can survive for extended periods in the damp, cool conditions in which fresh produce or shellfish (likely vehicles for *Giardia* cysts) are generally stored, and can also survive harsher conditions such as contact with chlorine, although they are susceptible to desiccation, heating and freeze-thawing, as well as some types of irradiation and chemical treatments. Nevertheless, the resilience of *Giardia* cysts to environmental pressures means that inactivation or decontamination of food products that are to be eaten with minimal processing in order to retain their sensory qualities, such as taste and texture, is probably the wrong approach; it is better to avoid contamination to begin with.

As *Giardia* cysts as contaminants of food products must originate from a human or an animal source, the most effective means of controlling such contamination from occurring on fresh produce is application of Good Agricultural Practice (GAP) during primary production, Good Manufacturing Practice (GMP) during processing and Good Hygienic Practice (GHP) before consumption (Dawson 2005). GAP includes using clean water for irrigation, fertiliser/pesticide application and washing, ensuring that domestic animals do not graze or contaminate horticultural areas, and taking precautions to ensure that wild animals do not have access to these growing areas. Wild animals excreting *Giardia* cysts with zoonotic potential have been detected in irrigation catchments and have been suggested to pose a potential threat to human health via crop contamination (McCarthy et al. 2008).

The pretreatment of wastewater in waste stabilisation ponds before using it as irrigation water was found to reduce contamination of crops with *Giardia* cysts in Morocco (Amahmid et al. 1999), while another study suggested that raw wastewater could be ozone treated for 1 h prior to use for agricultural purposes in order to inactivate *Giardia* cysts and other parasitic pathogens (Orta de Velásquez et al. 2006). A further study suggested that membrane ultrafiltration using a submerged hollow-fibre system could be suitable for treating wastewater prior to using it for irrigation to ensure removal of *Giardia* cyst contamination (Lonigro et al. 2006).

During processing procedures (e.g. washing, chopping, packaging), water of potable standard should also be used, and if the wash-water is reused then the

L.J. Robertson, *Giardia as a Foodborne Pathogen*, SpringerBriefs
in Food, Health, and Nutrition, DOI 10.1007/978-1-4614-7756-3_6,
© Lucy J. Robertson 2013

disinfection procedures should be effective at inactivating *Giardia* cysts (Cook and Lim 2012). It should be noted that the organic load in wash-water becomes very high (Rosenblum et al. 2012), and thus a disinfectant dose that is sufficient to inactivate *Giardia* cysts in clean water may be ineffective in wash-water.

During many food-manufacturing processes, elimination of microbial pathogens from foods is achieved by a variety of methods, including (most commonly) heat and chemical disinfection, and also irradiation or high pressure. Although the most commonly used sanitizer is traditionally chlorine (usually applied in the form of hypochlorous acid (HOCl), which is considered to be most effective), human health and environmental concerns (production of potentially carcinogenic by-products such as trihalomethanes and haloacetic acids and generation of wastewater with high levels of biological oxygen demand) have led some European authorities to prohibit the use of chlorine in organic produce washing. For example, several European countries, including Germany, Denmark, Holland and France, have banned the use of chlorine in washing organic produce, and alternatives, such as ozonisation or neutral electrolysed water, have been explored (Rosenblum et al. 2012; Abadias et al. 2008). Nevertheless, chlorination remains the most commonly used commercial sanitizing agent, with concentrations used for food application ranging from 50 to 200 ppm. For fresh produce, the most common application is 100 ppm hypochlorite; at pH 6.8–7.1, this dose yields 30–40 ppm free chlorine, depending upon organic load, for a contact time of 2 min at 4 °C, and is considered to be effective by affecting cell membrane proteins and inhibiting the activities of enzymes (cyst wall disruption, lysis of peripheral vacuoles, nuclear degradation and damage to the plasma lemma have been reported to be the main aspects of *Giardia* cyst injury from contact with chlorine; Li et al. 2004). Although contact with 1.5 ppm chlorine for less than 10 min at 25 °C has been reported to result in a 99 % reduction in the viability of *Giardia* cysts, 8 ppm is necessary to achieve the same effect at 5 °C (Jarroll et al. 1981), and thus the efficacy of standard chlorination at inactivating *Giardia* cysts on fresh produce is dependent on a range of factors. However, 30 ppm is considerably higher than the 8 ppm considered effective at 5 °C. For some more delicate produce, such as soft fruit (e.g. strawberries and raspberries), a quick spray with, or a brief (10 s) immersion in, 15–20 ppm free chlorine is used—depending on a range of factors, this may provide insufficient contact time for effective inactivation of *Giardia* cysts. The efficacy of sodium dichloroisocyanurate (NaDCC) at inactivating *Giardia* cysts on raw vegetables and fruits has been investigated (El Zawawy et al. 2010) and considered effective; the authors suggest that it may be suitable for use at the household and restaurant level, as well as in catering and fresh produce industry, mentioning its convenience in dry tablet format, and also its cheapness. Other technologies that may be useful in the fresh produce industry for inactivating *Giardia* cysts include UV irradiation, high-pressure processing, cobalt-60 irradiation (Sundermann and Estridge 2010) and, as mentioned previously, ozonisation. It should be noted that the use of sequential inactivation treatments might optimise existing treatments through synergistic effects (Erickson and Ortega 2006).

Producers have also considered the use of a range of "organic" products for decontaminating fresh produce. In particular, edible films of essential oils (including essential oils of oregano, allspice, cinnamon, thyme and clove bud) have been tested for their antimicrobial properties, with terpenoids and phenolic compounds considered to be largely responsible for their efficacy (Muriel-Galet et al. 2012a), along with films containing enzymes, and organic acids, such as lauric arginate (Muriel-Galet et al. 2012b). Although the efficacy of these approaches against protozoan parasites such as *Giardia* on fresh produce has not been tested, some experiments have considered the use of household sanitizers such as vinegar and lemon juice, and produced some results that are promising, and may be of use at the small-scale/market-garden level (Sadjjadi et al. 2006; Costa et al. 2009). Further experimentation on such substances may be of benefit in some situations.

In the shellfish industry, *Giardia* transmission is probably considered a relatively minor problem—if considered at all—and thus it may be useful to consider whether disinfection and inactivation treatments used for other more common treatments in the shellfish industry may be also effectual against *Giardia*. As well as depuration, other processes have been considered for shellfish such as oysters, largely because *Vibrio* species tend to be firmly attached to shellfish tissues and not readily removed in depuration tanks. Other treatments that have been considered for shellfish include high hydrostatic pressure, ultra-low temperature freezing, mild heat treatment, frozen dry storage, chemical treatments and irradiation (Wright et al. 2009). Unfortunately, it seems that those treatments that are effective against pathogens that are more commonly associated with shellfish tend also to kill the oyster, and thus introduce further problems regarding shelf life and storage. In addition, whether such treatments are effective against *Giardia* cysts in shellfish has not been explored.

Regardless of the type of food product, assessing inactivation effects of different treatment protocols on *Giardia* cyst viability requires appropriate methods for assessing the viability or the infectivity of the cysts. Unlike for bacteria, the cultivation of *Giardia in vitro* is frequently difficult; lack of establishment of an *in vitro* culture may be the result of a range of cultivation issues rather than pre-cultivation treatments, and animal model infectivity (using mice or gerbils) is also difficult for some isolates, as well as incorporating ethical issues. Use of morphological criteria (as observed by microscopy, preferably with DIC optics) along with inclusion or exclusion vital dyes provides an estimate of cyst viability, but is probably too imprecise for providing an industry standard and also frequently overestimates viability, and thus underestimates the efficacy of a treatment that is being investigated (Erickson and Ortega 2006). In assessing the efficacy of NADCC at inactivating *Giardia* cysts on fresh produce, El Zawawy et al. (2010) used *in vitro* excystation, trypan blue staining and bioassay in laboratory animals—however all these methods can be fraught with difficulties in interpretation. Molecular methods for assessing viability seem promising, but their lack of advance from the research lab to industry is indicative that they can also be problematic, and no single method for determining individual *Giardia* cyst viability and/or infectivity has been widely accepted to date. Probably these practical issues, as well as greater focus on bacterial (and, more recently, virus) pathogens on produce than on protozoal pathogens, means that there has been negligible research on the survival of *Giardia* cysts under the different sanitizing regimes used in the food industry.

Chapter 7
Risk Assessment and Regulations

The principal legislation that is in place to control the spread of *Giardia*, and similar protozoan parasites such as *Cryptosporidium*, in food within the manufacturing, processing, distribution, catering and retail sectors includes local Food Safety Acts or Regulations. Such Acts or Regulations tend not to have any international impact, and are directed towards specific perceived or identified problems. However, the foremost food safety management system for many years (since first devised to ensure that foods consumed by astronauts were safe) has been HACCP (supported by GHP and GMP), the intention of which is to provide a systematic, cost-effective and efficacious approach for risk management and prevention, and is of international application.

In addition, responsible authorities set public health targets that must be included by those involved in the provision of food. For example, the World Trade Organization introduced the concept of appropriate level of protection (ALOP) as a public health target, and other concepts have been introduced to enable these targets to be translated into meaningful, tangible objectives for the food industry. These concepts include Food Safety Objectives (FSOs), Performance Objectives (POs) and Performance Criteria (PC) proposed by the International Commission on Microbiological Specifications for Foods (ICMSF) and adopted by the Codex Alimentarius Food Hygiene Committee. However, the method by which FSOs can be extrapolated from or to ALOPs has not yet been established, although it is considered that FSOs provide a functional link between risk assessment and risk management, with HACCP acknowledged to be the principal tool available for use in the food industry.

While HACCP enables the identification of hazards and measures for their control, and the determination of critical control points along the farm/fjørd to fork continuum, for food producers, it differs from risk assessment. Risk assessment, comprising hazard identification, hazard characterization, exposure assessment and risk characterization, is a complementary tool to HACCP. In HACCP, multiple hazards for a single product in a particular facility are considered, whilst risk assessment traditionally focuses on single pathogen–food combinations. However, for

L.J. Robertson, *Giardia as a Foodborne Pathogen*, SpringerBriefs
in Food, Health, and Nutrition, DOI 10.1007/978-1-4614-7756-3_7,
© Lucy J. Robertson 2013

both HACCP and risk assessment, the key objective is risk mitigation. The HACCP system was originally targeted towards microbiological safety and is based on seven principles:

- Hazard analysis (identification of hazards and assessment of their severity and risk)
- Identification of critical control points (CCP)
- Specification of criteria to ensure control (establishment of critical limits)
- Monitoring of critical CCP (establishment of a system by which CCP can be monitored)
- Implementation of corrective action whenever monitoring indicates that criteria are not being met (establishment of appropriate corrective actions when a CCP is not under control)
- Verification that the system is functioning as planned (establishment of procedures for verification to confirm that HACCP is working effectively)
- Documentation for all procedures and records appropriate to these principles and their applications

These principles have similarities, and differences, to the four established stages in risk assessment:

- Hazard identification: Identification of agents capable of causing adverse health effects and that may be found in a particular food or food group
- Exposure assessment: Qualitative and/or quantitative evaluation of the likely intake of the relevant agents via food (also exposures from other sources if relevant)
- Hazard characterization: Qualitative and/or quantitative evaluation of the nature of the adverse health effects. For microbiological risk assessment the concerns relate directly to microorganisms (and/or, where relevant, their toxins)
- Risk characterization: Qualitative and/or quantitative estimation, including attendant uncertainties, of the probability of occurrence and severity of known or potential adverse health effects in a given population based on the three previous steps

Quantitative microbiological risk assessment (QMRA) is based on risk assessment as described above, but has the fundamental aims of protecting consumers, enabling decision-making on food safety issues, and assisting relevant authorities to meet public health goals by providing numerical limits and targets (within defined ranges). By focussing on risks that are associated with single hazards and product groups, it is possible to identify products of greatest concern to public health and those aspects of their processing/handling that impact most on risk. Thus, there is a clear overlap with, and support for, the HACCP motifs. While risk assessment in general, and QMRA in particular, can enhance HACCP by aiding in the identification of "design" CCPs, detailed considerations of specific facilities or locations are not a part of risk assessments that have a more general basis.

QMRAs are very data demanding, and, for *Giardia*, obtaining the data necessary to develop a robust QMRA is challenging. However, by reviewing the available published data and assessing it critically, it is possible to identify the gaps that must be filled before the QMRA can be conducted. Data can be acquired from different sources by retrieving from publications, expert elicitations, performing studies or developing models. More recently developed QMRAs tend to be stochastic in nature, with single values replaced by probability distributions that better reflect the uncertainty associated with inputs and derived parameters. It is obvious that the more relevant and accurate the data fed into a QMRA, the lower the uncertainty in risk estimates, and thus the more valuable the guidance for decision-making. Documentation and transparency are also key factors. Microbial risk assessment, particularly with application to risk from drinking water, first became widespread around 20 years ago, and QMRA is a commonly used tool in the water industry, with reference to *Giardia* as well as other pathogens. Within the water industry, QMRA is considered to be essential in the construction of Water Safety Plans (e.g. Smeets et al. 2010), to the extent that user-friendly online tools have been developed for water industries. In such tools, if the necessary data is loaded by an operator, who needs to have no special experience in QMRA-modelling, then a risk outcome associated with a particular pathogen, including *Giardia*, for consumption of drinking water from a particular source is obtained (Schijven et al. 2011). Such tools have not been widely implemented for irrigation water, although the principle would be the same, and one of the few published quantitative risk assessments considering protozoan parasites attempts to estimate the number of people that would be affected by fresh produce irrigated with water contaminated with protozoan parasites, including *Giardia* (Mota et al. 2009). In this study input into the QMRA included the following:

- The number of *Giardia* cysts per 100 l of irrigation water, with 58 surface water samples collected from widely dispersed points in one of the largest horticultural production areas of Mexico, and including samples taken from rivers, irrigation channels and drainage channels

  – Results ranged from 17 to 1,633 cysts per 100 l with a geometric mean of 82.34 cysts per 100 l

- The recovery efficiency of the detection method

  – 12 %

- The volume of irrigation water estimated to be retained on fresh produce

  – This varied according to produce type and was estimated to be 0.0036 ml/g for tomatoes, bell peppers and cucumbers, and 0.108 ml/g for lettuce (based on data from Shuval et al. 1997)

- The daily consumption of the different produce types by an adult in the USA

  – This varied according to produce type and was estimated to be 13.0 g (tomatoes); 4.3 g (bell peppers); 3.3 g (cucumbers); and 6.2 g (lettuce) (based on information from an online food consumption database run by the US Department of Agriculture)

For the purposes of the QMRA, a worst-case scenario was assumed, with 100 % transfer of the *Giardia* cysts from the irrigation water to the produce, and all the cysts being infectious to humans. Use of a "worst-case scenario" in such risk assessments, assuming, for example, raw consumption of produce, or overhead irrigation, is not unusual, particularly when data are limited (Hamilton et al. 2006). From this input, the estimated annual risks of infection (assuming 120 days exposure per year) with the *Giardia* cysts on the produce were found to range from 5.2×10^{-5}, associated with bell peppers contaminated via irrigation water with the lowest concentration of *Giardia* cysts (17 cysts per 100 l), up to as high as 1.96×10^{-1}, associated with lettuce contaminated via irrigation water with the highest concentration of *Giardia* cysts (1,633 cysts per 100 l) (Mota et al. 2009).

Despite limitations within a process such as this, it provides data that are useful for considering appropriate mitigation strategies/intervention practices or guidelines for pathogen reduction requirements. Other QMRAs have looked at specific processes, for example use of wastewater or sewage sludge, including for fertilising vegetable crops, and the potential for spread of pathogens, including *Giardia* (Westrell et al. 2004).

In a review of foodborne illness associated with *Cryptosporidium* and *Giardia* from livestock, Budu-Amoako et al. (2011) listed a range of interventions and mitigation strategies to reduce the contamination of fresh produce. These include on-farm interventions (GAP) to minimise infection in animals and further transmission of infection, watershed interventions to reduce contamination of water sources from parasites excreted from infected animals and food processing plant interventions to prevent food contamination, and with emphasis on the application of HACCP. Although the authors are concerned with prevention of contamination of the environment, particularly water sources, they apparently do not consider mixed farms, where animals and food crops may both be raised, and it is important that sufficient barriers are kept between the two types of agricultural commodities, both spatially and temporally, such that transfer from animals to plants via equipment or human transfer is minimised by ensuring that specific interventions (such as use of protective clothing or dedicated equipment) are instigated and followed.

Chapter 8
Future Challenges

Although we continually acquire more information and improved technologies that can be used to combat foodborne transmission of giardiasis, new challenges are being posed by a changing globe. Improved refrigeration for transport of fresh produce means that we now import or export our food as never before, producing complicated trade routes, more handling and greater possibilities for contamination, as well as potential difficulties for trace-back should contamination events occur. While improved food traceability systems have the potential to be used for some produce (e.g. radio-frequency identification—Kumar et al. 2009), the application of such technologies for third countries and small-scale producers is challenging and implementation will probably be driven by cost–benefit as well as consumer demand. Globalisation applies not only to food products, but also to the people who handle food along the farm-to-fork continuum and to the people who consume products— travel and tourism continue to flourish despite global economic downturns—and seasonal tasks such as crop picking may involve the use of an itinerant workforce.

Additionally, cultural changes and habits always seek to amaze us. In Japan, where ownership of pets is often forbidden in some apartments, animal cafés where clients can pet the animals before eating and drinking have become popular. The first cat café opened in Osaka in 2004, and apparently there are over 35 in Tokyo, and rabbit cafés, run on the same lines, are also popular. While there are strict rules to ensure hygiene and animal welfare, a study from 16 cat cafés in Tokyo reported Six *Giardia*-positive samples from 2 cat cafés belonging to the zoonotic genotype assemblage AI (Suzuki et al. 2011), and the authors suggest that, in association with eating and drinking, there is a risk of transmission from cats to humans.

Climate change is also a future challenge that may impact on contamination of fresh produce with *Giardia* cysts. Extreme weather events are known to result in overflow of sewage and the contamination of agricultural fields with faecal matter from both humans and animals.

L.J. Robertson, *Giardia as a Foodborne Pathogen*, SpringerBriefs
in Food, Health, and Nutrition, DOI 10.1007/978-1-4614-7756-3_8,
© Lucy J. Robertson 2013

Chapter 9
Conclusions

Although *Giardia* is generally considered as a parasitic infection of relatively minor severity, associated with low mortality and morbidity, clinical giardiasis nevertheless is also associated with prolonged and unpleasant symptoms, which may, in very rare cases, result in, or contribute to, death. In association with other insults to health, especially concomitant infections or compromised nutritional status, its effects may be particularly severe. Giardiasis may thus be considered a "neglected disease", both in countries with less advanced infrastructures but also in the most wealthy countries in the world; it should not be forgotten that a large outbreak of waterborne giardiasis made a considerable impact in Bergen, Norway, less than 10 years ago. The reason, in part, for the extensiveness of this outbreak was the lack of consideration by the medical and other authorities that giardiasis could be an autochthonous infection. Although *Giardia* is known as a pathogen that is transmitted via the faecal-oral route, it is seldom particularly associated with food as an infection vehicle. This is probably because there are relatively few outbreaks of giardiasis for which a food vehicle has been positively identified. Nevertheless, the potential for foodborne infection is evident; the transmission stage is resilient to environmental pressures, the infectious dose is low and *Giardia* cysts are excreted in enormous quantities by infected hosts. Thus, probably the reason that infections are seldom recognised as foodborne is not because foodborne transmission does not occur, but because clinical, diagnostic and epidemiological barriers hamper the ease of making the correct associations; the symptoms start several days after the implicated food has been eaten, the clinician may fail to make the correct diagnosis and the resultant gap of days—maybe weeks or even months—between consumption of contaminated food and diagnosis of giardiasis frustrates epidemiological investigation. Thus foodborne giardiasis probably occurs considerably more often than would be expected by direct extrapolation from the literature.

How food becomes contaminated, which food products are most likely to be contaminated and how *Giardia* cysts that have contaminated food can be removed and/or inactivated are all questions about which the data are relatively limited.

L.J. Robertson, *Giardia as a Foodborne Pathogen*, SpringerBriefs
in Food, Health, and Nutrition, DOI 10.1007/978-1-4614-7756-3_9,
© Lucy J. Robertson 2013

Nevertheless, some surveys have been conducted and these indicate that fresh produce (in particular), shellfish and beverages (to a lesser extent) and possibly meat can act as potential food vehicles for transmission. These data have been the driving forces for developing a standardised method for examination of food (especially fresh produce—leafy greens and small red fruits) for contamination with *Giardia* cysts, and an ISO Method, based on elution, concentration, isolation and detection, is under development (registered in the ISO/TC34/SC9 work programme with the number ISO 18744). The use of molecular methods to determine whether *Giardia* cysts are of a genotype that is of public health significance can be added onto the method subsequently. While the information obtained using such methods is valuable for increasing our understanding of foodborne transmission of giardiasis, there are also obvious gaps in our knowledge. These include not only our understanding of the pathogenicity of *Giardia*—why do infections in some people cause little or no symptoms, while in others giardiasis can be a prolonged and unpleasant illness?—but also we require better methods for evaluating *Giardia* cyst viability and infectivity and thereby developing appropriate methods for removal or inactivation of *Giardia* cysts along the farm-to-fork continuum. In the absence of effective methods for this, the rigorous use of HACCP and risk analysis in order to reduce contamination and optimise appropriate interventions that will minimise transmission risk are very important.

References

Abadias M, Usall J, Oliveira M, Alegre I, Viñas I (2008) Efficacy of neutral electrolyzed water (NEW) for reducing microbial contamination on minimally-processed vegetables. Int J Food Microbiol 123(1–2):151–158. doi:10.1016/j.ijfoodmicro.2007.12.008

Adak G, Long S, O'Brien S (2002) Trends in indigenous foodborne disease and deaths, England and Wales: 1992 to 2000. Gut 51:832–841

Amahmid O, Asmama S, Bouhoum K (1999) The effect of wastewater reuse in irrigation on the contamination level of food crops by *Giardia* cysts and *Ascaris* eggs. Int J Food Microbiol 49:19–26

Amorós I, Alonso JL, Cuesta G (2010) *Cryptosporidium* oocysts and *Giardia* cysts on salad products irrigated with contaminated water. J Food Prot 73:1138–1140

Andargic G, Kassu A, Moges F, Tiruneh M, Huruy K (2008) Prevalence of bacteria and intestinal parasites among food-handlers in Gondar town, northwest Ethiopia. J Health Popul Nutr 26(4):451–455

Babiker MA, Ali MS, Ahmed ES (2009) Frequency of intestinal parasites among food-handlers in Khartoum, Sudan. East Mediterr Health J 15(5):1098–1104

Baldursson S, Karanis P (2011) Waterborne transmission of protozoan parasites: review of worldwide outbreaks—an update 2004–2010. Water Res 45(20):6603–6614. doi:10.1016/j.watres.2011.10.013

Beck R, Sprong H, Bata I, Lucinger S, Pozio E, Caccio SM (2011) Prevalence and molecular typing of Giardia spp. in captive mammals at the zoo of Zagreb, Croatia. Vet Parasitol 175:40–46. doi:10.1016/j.vetpar.2010.09.026

Bello J, Núñez FA, González OM, Fernández R, Almirall P, Escobedo A (2011) Risk factors for *Giardia* infection among hospitalized children in Cuba. Ann Trop Med Parasitol 105(1):57–64. doi:10.1179/136485911X12899838413385

Benchimol K, De Souza W (2011) The ultrastructure of Giardia during growth and differentiation. In: Svärd S, Luján HD (eds) *Giardia*: a model organism. Springer, New York. ISBN 978-3-7091-0197-1

Bénéré E, Van Assche T, Cos P, Maes L (2011) Variation in growth and drug susceptibility among Giardia duodenalis assemblages A, B and E in axenic in vitro culture and in the gerbil model. Parasitology 138(11):1354–1361. doi:10.1017/S0031182011001223

Bénéré E, Van Assche T, Van Ginneken C, Peulen O, Cos P, Maes L (2012) Intestinal growth and pathology of Giardia duodenalis assemblage subtype A(I), A(II), B and E in the gerbil model. Parasitology 139(4):424–433. doi:10.1017/S0031182011002137

Berkman DS, Lescano AG, Gilman RH, Lopez SL, Black MM (2002) Effects of stunting, diarrhoeal disease, and parasitic infection during infancy on cognition in late childhood: a follow-up study. Lancet 59(9306):564–571

Birky CW Jr (2010) Giardia sex? Yes, but how and how much? Trends Parasitol 26:70–74. doi:10.1016/j.pt.2009.11.007

Budu-Amoako E, Greenwood SJ, Dixon BR, Barkema HW, McClure JT (2011) Foodborne illness associated with Cryptosporidium and Giardia from livestock. J Food Prot 74(11):1944–1955. doi:10.4315/0362-028X.JFP-11-107

Buret AG, Cotton J (2011) Pathophysiological processes and clinical manifestations of giardiasis. In: Luján HD, Svärd S (eds) Giardia: a model organism. Springer, New York. ISBN 978-3-7091-0197-1

Cacciò SM, Sprong H (2011) Epidemiology of giardiasis in humans. In: Svärd S, Luján HD (eds) Giardia: a model organism. Springer, New York. ISBN 978-3-7091-0197-1

Cann KF, Thomas DR, Salmon RL, Wyn-Jones AP, Kay D (2013) Extreme water-related weather events and waterborne disease. Epidemiol Infect 141(4):671–686. doi:10.1017/S0950268812001653

Cantey PT, Roy S, Lee B, Cronquist A, Smith K, Liang J, Beach MJ (2011) Study of non-outbreak giardiasis: novel findings and implications for research. Am J Med 124(12):1175.e1–1175.e8. doi:10.1016/j.amjmed.2011.06.012

Casteel MJ, Sobsey MD, Mueller JP (2006) Fecal contamination of agricultural soils before and after hurricane-associated flooding in North Carolina. J Environ Sci Health A Tox Hazard Subst Environ Eng 41(2):173–184

Chaidez C, Soto M, Gortares P, Mena K (2005) Occurrence of Cryptosporidium and Giardia in irrigation water and its impact on the fresh produce industry. Int J Environ Health Res 15:339–345

Conn DB, Weaver J, Tamang L, Graczyk TK (2007) Synanthropic flies as vectors of Cryptosporidium and Giardia among livestock and wildlife in a multispecies agricultural complex. Vector Borne Zoonotic Dis 7(4):643–651

Conroy DA (1960) A note on the occurrence of Giardia sp. in a Christmas pudding. Rev Iber Parasitol 20:567–571

Cook N, Lim YAL (2012) Giardia duodenalis: contamination of fresh produce. In: Robertson LJ, Smith HV (eds) Foodborne protozoan parasite. Nova Publishers, Hauppauge, NY

Cook N, Nichols RAB, Wilkinson N, Paton CA, Barker K, Smith HV (2007) Development of a method for the detection of Giardia duodenalis on lettuce and for simultaneous analysis of salad products from the presence of Giardia cysts and Cryptosporidium oocysts. Appl Environ Microbiol 73:7388–7391

Costa AO, Thomaz-Soccol V, Paulino RC, Alcântara de Castro E (2009) Effect of vinegar on the viability of Giardia duodenalis cysts. Int J Food Microbiol 128(3):510–512

Costa-Cruz JM, Cardoso ML, Marques DE (1995) Intestinal parasites in school food handlers in the city of Uberlândia, Minas Gerais, Brazil. Rev Inst Med Trop Sao Paulo 37(3):191–196

Dawson D (2005) Foodborne protozoan parasites. Int J Food Microbiol 103:207–227

de Leon WU, Monzon RB, Aganon AA, Arceo RE, Ignacio EJ, Santos G (1992) Parasitic contamination of selected vegetables sold in Metropolitan Manila, Philippines. Southeast Asian J Trop Med Public Health 23(1):162–164

DeRegnier DP, Cole L, Schupp DG, Erlandsen SL (1989) Viability of Giardia cysts suspended in lake, river, and tap water. Appl Environ Microbiol 55(5):1223–1229

Di Benedetto MA, Cannova L, Di Piazza F, Amodio E, Bono F, Cerame G, Romano N (2007) Hygienic-sanitary quality of ready-to-eat salad vegetables on sale in the city of Palermo (Sicily). Ig Sanita Pubbl 63:659–670

Diallo MB, Anceno AJ, Tawatsupa B, Tripathi NK, Wangsuphachart V, Shipin OV (2009) GIS-based analysis of the fate of waste-related pathogens Cryptosporidium parvum, Giardia lamblia and Escherichia coli in a tropical canal network. J Water Health 7(1):133–143. doi:10.2166/wh.2009.010

Dib HH, Lu SQ, Wen SF (2008) Prevalence of Giardia lamblia with or without diarrhea in South East, South East Asia and the Far East. Parasitol Res 103(2):239–251. doi:10.1007/s00436-008-0968-6

Dixon BR (2009) The role of livestock in the foodborne transmission of *Giardia duodenalis* and *Cryptosporidium* spp. to humans. In: Ortega-Pierres MG, Caccio SM, Fayer R, Smith H (eds) *Giardia* and *Cryptosporidium*: from molecules to disease. CAB International, Wallingford, UK, pp 107–122

Dixon B, Parrington L, Cook A, Pollari F, Farber J (2013) Detection of *Cyclospora*, *Cryptosporidium*, and *Giardia* in Ready-to-Eat packaged leafy greens in Ontario, Canada. J Food Prot 76(2):307–313. doi:10.4315/0362-028X.JFP-12-282

El Zawawy LA, El-Said D, Ali SM, Fathy FM (2010) Disinfection efficacy of sodium dichloroisocyanurate (NADCC) against common food-borne intestinal protozoa. J Egypt Soc Parasitol 40(1):165–185

Ensink JH, van der Hoek W (2009) Implementation of the WHO guidelines for the safe use of wastewater in Pakistan: balancing risks and benefits. J Water Health 7(3):464–468. doi:10.2166/wh.2009.061

Ensink JH, van der Hoek W, Amerasinghe FP (2006) *Giardia duodenalis* infection and wastewater irrigation in Pakistan. Trans R Soc Trop Med Hyg 100(6):538–542

Ensink JH, Mahmood T, Dalsgaard A (2007) Wastewater-irrigated vegetables: market handling versus irrigation water quality. Trop Med Int Health 12(Suppl 2):2–7

Erickson MC, Ortega YR (2006) Inactivation of protozoan parasites in food, water, and environmental systems. J Food Prot 69(11):2786–2808

Escobedo AA, Almirall P, Robertson LJ, Franco RM, Hanevik K, Mørch K, Cimerman S (2010) Giardiasis: the ever-present threat of a neglected disease. Infect Disord Drug Targets 10(5):329–348

Espelage W, An der Heiden M, Stark K, Alpers K (2010) Characteristics and risk factors for symptomatic *Giardia lamblia* infections in Germany. BMC Public Health 10:41. doi:10.1186/1471-2458-10-41

Fallaha AA, Pirali-Kheirabadib K, Shirvanid F, Saei-Dehkordi SS (2012) Prevalence of parasitic contamination in vegetables used for raw consumption in Shahrekord, Iran: influence of season and washing procedure. Food Control 25:617–620

Fayer R, Graczyk TK, Lewis EJ, Trout JM, Farley CA (1998) Survival of infectious *Cryptosporidium parvum* oocysts in seawater and eastern oysters (*Crassostrea virginica*) in the Chesapeake Bay. Appl Environ Microbiol 64:1070–1074

Fernando WJ (2009) Theoretical considerations and modeling of chemical inactivation of microorganisms: inactivation of *Giardia* cysts by free chlorine. J Theor Biol 259(2):297–303. doi:10.1016/j.jtbi.2009.03.026

Freites A, Colmenares D, Pérez M, García M, Díaz de Suárez O (2009) *Cryptosporidium* sp. infections and other intestinal parasites in food handlers from Zulia state, Venezuela. Invest Clin 50(1):13–21

Frenzen PD (2004) Deaths due to unknown foodborne agents. Emerg Infect Dis 10:1536–1543

Gerwig GJ, van Kuik JA, Leeflang BR, Kamerling JP, Vliegenthart JF, Karr CD, Jarroll EL (2002) The *Giardia intestinalis* filamentous cyst wall contains a novel beta(1-3)-*N*-acetyl-D-GALACTOSAMINE polymer: a structural and conformational study. Glycobiology 12(8):499–505

Geurden T, Olsen M (2011) Giardia in pets and farm animals, and their zoonotic potential. In: Svärd S, Luján HD (eds) *Giardia*: a model organism. Springer, New York. ISBN 978-3-7091-0197-1

Gil MI, Selma MV, López-Gálvez F, Allende A (2009) Fresh-cut product sanitation and wash water disinfection: problems and solutions. Int J Food Microbiol 134(1–2):37 45. doi:10.1016/j.ijfoodmicro.2009.05.021

Girotto KG, Grama DF, Cunha MJ, Faria ES, Limongi JE, Pinto Rde M, Cury MC (2013) Prevalence and risk factors for intestinal protozoa infection in elderly residents at Long Term Residency Institutions in Southeastern Brazil. Rev Inst Med Trop Sao Paulo 55(1):19–24

Gómez-Couso H, Ares-Mazás ME (2012) *Giardia duodenalis*: contamination of bivalve molluscs. In: Robertson LJ, Smith HV (eds) Foodborne protozoan parasite. Nova Publishers, Hauppauge, NY

Gómez-Couso H, Freire-Santos F, Amar CF, Grant KA, Williamson K, Ares-Mazás ME, McLauchlin J (2004) Detection of *Cryptosporidium* and *Giardia* in molluscan shellfish by multiplexed nested-PCR. Int J Food Microbiol 91:279–288

Gómez-Couso H, Méndez-Hermida F, Castro-Hermida JA, Ares-Mazás E (2005a) *Giardia* in shellfish-farming areas: detection in mussels, river water and waste waters. Vet Parasitol 133:13–18

Gómez-Couso H, Méndez-Hermida F, Castro-Hermida JA, Ares-Mazás E (2005b) Occurrence of *Giardia* cysts in mussels (*Mytilus galloprovincialis*) destined for human consumption. J Food Prot 68:1702–1705

Gortáres-Moroyoqui P, Castro-Espinoza L, Naranjo JE, Karpiscak MM, Freitas RJ, Gerba CP (2011) Microbiological water quality in a large irrigation system: El Valle del Yaqui, Sonora México. J Environ Sci Health A Tox Hazard Subst Environ Eng 46(14):1708–1712. doi:10.10 80/10934529.2011.623968

Graczyk TK, Conn DB, Marcogliese DJ, Graczyk H, De Lafontaine Y (2003) Accumulation of human waterborne parasites by zebra mussels (*Dreissena polymorpha*) and Asian freshwater clams (*Corbicula fluminea*). Parasitol Res 89(2):107–112

Graczyk TK, Girouard AS, Tamang L, Nappier SP, Schwab KJ (2006) Recovery, bioaccumulation, and inactivation of human waterborne pathogens by the Chesapeake Bay nonnative oyster, *Crassostrea ariakensis*. Appl Environ Microbiol 72(5):3390–3395

Greig JD, Todd EC, Bartleson CA, Michaels BS (2007) Outbreaks where food workers have been implicated in the spread of foodborne disease. Part 1. Description of the problem, methods, and agents involved. J Food Prot 70(7):1752–1761

Gündüz T, Limoncu ME, Cümen S, Ari A, Serdağ E, Tay Z (2008) The prevalence of intestinal parasites and nasal *S. aureus* carriage among food handlers. J Environ Health 70(10):64–65

Hamilton AJ, Stagnitti F, Premier R, Boland AM, Hale G (2006) Quantitative microbial risk assessment models for consumption of raw vegetables irrigated with reclaimed water. Appl Environ Microbiol 72(5):3284–3290

Hanevik K, Hausken T, Morken MH, Strand EA, Mørch K, Coll P, Helgeland L, Langeland N (2007) Persisting symptoms and duodenal inflammation related to *Giardia duodenalis* infection. J Infect 55(6):524–530

Jarroll EA, Bingham AK, Meyer EA (1981) Effect of chlorine on *Giardia lamblia* cyst viability. Appl Environ Microbiol 41:483–487

Johannessen G, Robertson L, Myrmel M, Jensvoll L (2013) Sluttrapport: Smittestoffer i vegetabilske næringsmidler. Veterinærinstituttets rapportserie 7 – 2013. ISSN: 1890-3290

Kamau P, Aloo-Obudho P, Kabiru E, Ombacho K, Langat B, Mucheru O, Ireri L (2012) Prevalence of intestinal parasitic infections in certified food-handlers working in food establishments in the City of Nairobi, Kenya. J Biomed Res 26(2):84–89. doi:10.1016/S1674-8301(12)60016-5

Karabiber N, Aktas F (1991) Foodborne giardiasis. Lancet 377:376–377

Karanis P, Kourenti C, Smith H (2007) Waterborne transmission of protozoan parasites: a worldwide review of outbreaks and lessons learnt. J Water Health 5:1–38

Keserue HA, Füchslin HP, Wittwer M, Nguyen-Viet H, Nguyen TT, Surinkul N, Koottatep T, Schürch N, Egli T (2012) Comparison of rapid methods for detection of *Giardia* spp. and *Cryptosporidium* spp. (oo)cysts using transportable instrumentation in a field deployment. Environ Sci Technol 46(16):8952–8959. doi:10.1021/es301974m

Khurana S, Taneja N, Thapar R, Sharma M, Malla N (2008) Intestinal bacterial and parasitic infections among food handlers in a tertiary care hospital of North India. Trop Gastroenterol 29(4):207–209

Kistemann T, Classen T, Koch C, Dangendorf F, Fischeder R, Gebel J, Vacata V, Exner M (2002) Microbial load of drinking water reservoir tributaries during extreme rainfall and runoff. Appl Environ Microbiol 68(5):2188–2197

Kniel KE, Lindsay DS, Sumner SS, Hackney CR, Pierson MD, Dubey JP (2002) Examination of attachment and survival of *Toxoplasma gondii* oocysts on raspberries and blueberries. J Parasitol 88(4):790–793

Koh W, Geurden T, Paget T, O'Handley R, Steuart R, Thompson A, Buret AG (2013) *Giardia duodenalis* assemblage-specific induction of apoptosis and tight junction disruption in human intestinal epithelial cells: effects of mixed infections. J Parasitol 99(2):353–358. doi:10.1645/GE-3021.1

Kumar P, Reinitz HW, Simunovic J, Sandeep KP, Franzon PD (2009) Overview of RFID technology and its applications in the food industry. J Food Sci 74(8):R101–R106. doi:10.1111/j.1750-3841.2009.01323.x

Kutz SJ, Thompson RCA, Polley L (2009) Wildlife with *Giardia*: villain, or victim and vector? In: Cacciò S, Fayer R, Mank T, Thompson RCA, Ortega-Pierres GS (eds) *Giardia* and *Cryptosporidium*: from molecules to disease. CABI Publishing, Wallingford, UK

Lalle M, Pozio E, Capelli G, Bruschi F, Crotti D, Cacciò SM (2005) Genetic heterogeneity at the beta-giardin locus among human and animal isolates of *Giardia duodenalis* and identification of potentially zoonotic subgenotypes. Int J Parasitol 35(2):207–213. doi:10.1016/j.ijpara.2004.10.022

Lane S, Lloyd D (2002) Current trends in research into the waterborne parasite *Giardia*. Crit Rev Microbiol 28(2):123–147

Lasek-Nesselquist E, Welch DM, Sogin ML (2010) The identification of a new *Giardia duodenalis* assemblage in marine vertebrates and a preliminary analysis of *G. duodenalis* population biology in marine systems. Int J Parasitol 40(9):1063–1074. doi:10.1016/j.ijpara.2010.02.015

Leal Diego AG, Dores Ramos AP, Marques Souza DS, Durigan M, Greinert-Goulart JA, Moresco V, Amstutz RC, Micoli AH, Neto RC, Monte Barardi CR, Bueno Franco RM (2013) Sanitary quality of edible bivalve mollusks in Southeastern Brazil using an UV based depuration system. Ocean Coast Manag 72:93–100. doi:10.1016/j.ocecoaman.2011.07.010

Lévesque B, Gagnon F, Valentin A, Cartier JF, Chevalier P, Cardinal P, Cantin P, Gingras S (2006) A study to assess the microbial contamination of *Mya arenaria* clams from the north shore of the St Lawrence River estuary, (Québec, Canada). Can J Microbiol 52(10):984–991. doi:10.1139/w06-061

Lévesque B, Barthe C, Dixon BR, Parrington LJ, Martin D, Doidge B, Proulx JF, Murphy D (2010) Microbiological quality of blue mussels (*Mytilus edulis*) in Nunavik, Quebec: a pilot study. Can J Microbiol 56(11):968–977. doi:10.1139/w10-078

Li YN, Smith DW, Belosevic M (2004) Morphological changes of *Giardia lamblia* cysts after treatment with ozone and chlorine. J Environ Eng Sci 3:495–506

Lonigro A, Pollice A, Spinelli R, Berrilli F, Di Cave D, D'Orazi C, Cavallo P, Brandonisio O (2006) *Giardia* cysts and *Cryptosporidium* oocysts in membrane-filtered municipal wastewater used for irrigation. Appl Environ Microbiol 72(12):7916–7918

Lucy FE, Graczyk TK, Tamang L, Miraflor A, Minchin D (2008) Biomonitoring of surface and coastal water for *Cryptosporidium*, *Giardia*, and human-virulent microsporidia using molluscan shellfish. Parasitol Res 103(6):1369–1375. doi:10.1007/s00436-008-1143-9

McCarthy S, Ng J, Gordon C, Miller R, Wyber A, Ryan UM (2008) Prevalence of *Cryptosporidium* and *Giardia* species in animals in irrigation catchments in the southwest of Australia. Exp Parasitol 118(4):596–599

Mead PS, Slutsker L, Dietz V, McCaig LF, Bresee JS, Shapiro C, Griffin PM, Tauxe RV (1999) Food-related illness and death in the United States. Emerg Infect Dis 5:607–625

Mintz ED, Hudson-Wragg M, Mshar P, Cartter ML, Hadler JL (1993) Foodborne giardiasis in a corporate office setting. J Infect Dis 167:250–253

Mohammed Mahdy AK, Lim YA, Surin J, Wan KL, Al-Mekhlafi MS (2008) Risk factors for endemic giardiasis: highlighting the possible association of contaminated water and food. Trans R Soc Trop Med Hyg 102(5):465–470. doi:10.1016/j.trstmh.2008.02.004

Monge R, Arias ML (1996) Presence of various pathogenic microorganisms in fresh vegetables in Costa Rica. Arch Latinoam Nutr 46:292–294

Monis PT, Cacciò SM, Thompson RC (2009) Variation in *Giardia*: towards a taxonomic revision of the genus. Trends Parasitol 25:93–100. doi:10.1016/j.pt.2008.11.006

Mørch K, Hanevik K, Robertson LJ, Strand EA, Langeland N (2008) Treatment-ladder and genetic characterisation of parasites in refractory giardiasis after an outbreak in Norway. J Infect 56(4):268–273. doi:10.1016/j.jinf.2008.01.013

Mossallam SF (2010) Detection of some intestinal protozoa in commercial fresh juices. J Egypt Soc Parasitol 40(1):135–149

Mota A, Mena KD, Soto-Beltran M, Tarwater PM, Cháidez C (2009) Risk assessment of *Cryptosporidium* and *Giardia* in water irrigating fresh produce in Mexico. J Food Prot 72(10):2184–2188

Muriel-Galet V, Cerisuelo JP, López-Carballo G, Lara M, Gavara R, Hernández-Muñoz P (2012a) Development of antimicrobial films for microbiological control of packaged salad. Int J Food Microbiol 157(2):195–201

Muriel-Galet V, López-Carballo G, Gavara R, Hernández-Muñoz P (2012b) Antimicrobial food packaging film based on the release of LAE from EVOH. Int J Food Microbiol 157(2):239–244

Nappier SP, Graczyk TK, Tamang L, Schwab KJ (2010) Co-localized *Crassostrea virginica* and *Crassostrea ariakensis* oysters differ in bioaccumulation, retention and depuration of microbial indicators and human enteropathogens. J Appl Microbiol 108(2):736–744. doi:10.1111/j.1365-2672.2009.04480.x

Nash TE (2011) Antigenic variation in *Giardia*. In: Svärd S, Luján HD (eds) *Giardia*: a model organism. Springer, New York. ISBN 978-3-7091-0197-1

Nash TE, Herrington DA, Losonsky GA, Levine MM (1987) Experimental human infections with *Giardia lamblia*. J Infect Dis 156:974–984

Negm AY (2003) Human pathogenic protozoa in bivalves collected from local markets in Alexandria. J Egypt Soc Parasitol 33:991–998

Nuñez FA, Robertson LJ (2012) Trends in foodborne diseases, including deaths. In: Robertson LJ, Smith HV (eds). Foodborne protozoan parasite. Nova Publishers, Hauppauge, NY

Orta de Velásquez MT, Rojas-Valencia MN, Reales-Pineda AC (2006) Evaluation of phytotoxic elements, trace elements and nutrients in a standardized crop plant, irrigated with raw wastewater treated by APT and ozone. Water Sci Technol 54(11–12):165–173

Osterholm MT, Forfang JC, Ristinen TL, Dean AG, Washburn JW, Godes JR, Rude RA, McCullough JG (1981) An outbreak of foodborne giardiasis. N Engl J Med 304:24–28

Petersen LR, Cartter ML, Hadler JL (1988) A food-borne outbreak of *Giardia lamblia*. J Infect Dis 157:846–848

Porter JD, Gaffney C, Heymann D, Parkin W (1990) Foodborne outbreak of *Giardia lamblia*. Am J Public Health 80:1259–1260

Quick R, Paugh K, Addiss D, Kobayashi J, Baron R (1992) Restaurant associated outbreak of giardiasis. J Infect Dis 166:673–676

Rai AK, Chakravorty R, Paul J (2008) Detection of *Giardia*, *Entamoeba*, and *Cryptosporidium* in unprocessed food items from northern India. World J Microbiol Biotechnol 24(12):2879–2887

Rendtorff RC (1954) The experimental transmission of human intestinal protozoan parasites. II. Giardia lamblia cysts given in capsules. Am J Hyg 59(2):209–220

Robertson LJ (2007) The potential for marine bivalve shellfish to act as transmission vehicles for outbreaks of protozoan infections in humans: a review. Int J Food Microbiol 120(3):201–216

Robertson LJ (2009) *Giardia* and *Cryptosporidium* infections in sheep and goats: a review of the potential for transmission to humans via environmental contamination. Epidemiol Infect 137(7):913–921. doi:10.1017/S0950268809002295

Robertson LJ (2013) Chapter 13: Protozoan parasites: a plethora of potentially foodborne pathogens. In: Tham W, Danielsson-Tham M-L (eds) Food associated pathogens. CRC, Boca Raton, FL. ISBN 978-1466584983

Robertson LJ, Chalmers RM (2013) Foodborne cryptosporidiosis: is there really more in Nordic countries? Trends Parasitol 29(1):3–9. doi:10.1016/j.pt.2012.10.003

Robertson LJ, Gjerde B (2000) Isolation and enumeration of *Giardia* cysts, *Cryptosporidium* oocysts, and *Ascaris* eggs from fruits and vegetables. J Food Prot 63:775–778

Robertson LJ, Gjerde BK (2001a) Occurrence of parasites on fruits and vegetables in Norway. J Food Prot 64:1793–1798

Robertson LJ, Gjerde B (2001b) Factors affecting recovery efficiency in isolation of *Cryptosporidium* oocysts and *Giardia* cysts from vegetables for standard method development. J Food Prot 64(11):1799–1805

Robertson LJ, Gjerde BK (2006) Fate of *Cryptosporidium* oocysts and *Giardia* cysts in the Norwegian aquatic environment over winter. Microb Ecol 52:597–602. doi:10.1007/s00248-006-9005-4

Robertson LJ, Gjerde B (2008) Development and use of a pepsin digestion method for analysis of shellfish for *Cryptosporidium* oocysts and *Giardia* cysts. J Food Prot 71(5):959–966

Robertson LJ, Lim YAL (2011) Waterborne and environmentally-borne giardiasis. In: Svärd S, Luján HD (eds) *Giardia*: a model organism. Springer, New York. ISBN 978-3-7091-0197-1

Robertson LJ, Hermansen L, Gjerde BK, Strand E, Alvsvåg JO, Langeland N (2006) Application of genotyping during an extensive outbreak of waterborne giardiasis in Bergen, Norway, during autumn and winter 2004. Appl Environ Microbiol 72:2212–2217. doi:10.1128/AEM.72.3.2212-2217.2006

Robertson LJ, Forberg T, Hermansen L, Gjerde BK, Langeland N (2008) Demographics of *Giardia* infections in Bergen, Norway, subsequent to a waterborne outbreak. Scand J Infect Dis 40(2):189–192. doi:10.1080/00365540701558672

Robertson LJ, Hanevik K, Escobedo AA, Mørch K, Langeland N (2010) Giardiasis—why do the symptoms sometimes never stop? Trends Parasitol 26:75–82. doi:10.1016/j.pt.2009.11.010

Robertson LJ, van der Giessen JWB, Batz MB, Kojima M, Cahill S (2013) Have foodborne parasites finally become a global concern? Trends Parasitol 29(3):101–103. doi:10.1016/j.pt.2012.12.004

Rosenblum J, Ge C, Bohrerova Z, Yousef A, Lee J (2012) Ozonation as a clean technology for fresh produce industry and environment: sanitizer efficiency and wastewater quality. J Appl Microbiol 113(4):837–845. doi:10.1111/j.1365-2672.2012.05393.x

Sadjjadi SM, Rostami J, Azadbakht M (2006) Giardiacidal activity of lemon juice, vinifer and vinegar on *Giardia intestinalis* cysts. Southeast Asian J Trop Med Public Health 37(Suppl 3):24–27

Savioli L, Smith H, Thompson A (2006) *Giardia* and *Cryptosporidium* join the 'Neglected Diseases Initiative'. Trends Parasitol 22(5):203–208

Scallan E, Hoekstra RM, Angulo FJ, Tauxe RV, Widdowson MA, Roy SL, Jones JL, Griffin PM (2011) Foodborne illness acquired in the United States—major pathogens. Emerg Infect Dis 17:7–15. doi:10.3201/eid1701.091101p1

Schets FM, van den Berg HH, Engels GB, Lodder WJ, de Roda Husman AM (2007) *Cryptosporidium* and *Giardia* in commercial and non-commercial oysters (*Crassostrea gigas*) and water from the Oosterschelde, The Netherlands. Int J Food Microbiol 113:189–194

Schijven JF, Teunis PF, Rutjes SA, Bouwknegt M, de Roda Husman AM (2011) QMRAspot: a tool for quantitative microbial risk assessment from surface water to potable water. Water Res 45(17):5564–5576. doi:10.1016/j.watres.2011.08.024

Selma MV, Allende A, López-Gálvez F, Conesa MA, Gil MI (2008a) Heterogeneous photocatalytic disinfection of wash waters from the fresh-cut vegetable industry. J Food Prot 71(2):286–292

Selma MV, Allende A, López-Gálvez F, Conesa MA, Gil MI (2008b) Disinfection potential of ozone, ultraviolet-C and their combination in wash water for the fresh-cut vegetable industry. Food Microbiol 25(6):809–814. doi:10.1016/j.fm.2008.04.005

Shahnazi M, Jafari-Sabet M (2010) Prevalence of parasitic contamination of raw vegetables in villages of Qazvin Province, Iran. Foodborne Pathog Dis 7(9):1025–1030. doi:10.1089/fpd.2009.0477

Shapiro K, Miller WA, Silver MW, Odagiri M, Largier JL, Conrad PA, Mazet JA (2013) Research commentary: association of zoonotic pathogens with fresh, estuarine, and marine macroaggregates. Microb Ecol 65(4):928–933. doi:10.1007/s00248-012-0147-2

Shuval H, Lampert Y, Fattal B (1997) Development of a risk assessment approach for evaluating wastewater reuse standards for agriculture. Water Sci Technol 35(11–12):15–20. doi:10.1016/S0273-1223(97)00228-X

Siala E, Guidara R, Ben Abdallah R, Ben Ayed S, Ben Alaya N, Zallaga N, Bouratbine A, Aoun K (2011) The intestinal parasites in the food handlers of Tunis area: study of 8502 stool samples (1998–2008). Arch Inst Pasteur Tunis 88(1–4):77–84

Singer SM (2011) Immunology of giardiasis. In: Svärd S, Luján HD (eds) Giardia: a model organism. Springer, New York. ISBN 978-3-7091-0197-1

Smeets PW, Rietveld LC, van Dijk JC, Medema GJ (2010) Practical applications of quantitative microbial risk assessment (QMRA) for water safety plans. Water Sci Technol 61(6): 1561–1568. doi:10.2166/wst.2010.839

Smith HV, Mank TG (2011) Diagnosis of human giardiasis. In: Svärd S, Luján HD (eds) Giardia: a model organism. Springer, New York. ISBN 978-3-7091-0197-1

Smith HV, Robertson LJ, Ongerth JE (1995) Cryptosporidiosis and giardiasis: the impact of waterborne transmission. J Water SRT-Aqua 44(6):258–274

Smith-DeWaal C, Barlow K, Alderton L, Jacobson MF (2001) Outbreak alert. Center for Science in the Public Interest, Washington, DC, 2001. Available at: http://www.cspinet.org/reports/oa_2001.pdf

Sprong H, Cacciò SM, van der Giessen JW, ZOOPNET network and partners (2009) Identification of zoonotic genotypes of Giardia duodenalis. PLoS Negl Trop Dis 3:e558. doi:10.1371/journal.pntd.0000558

Srikanth R, Naik D (2004) Prevalence of giardiasis due to wastewater reuse for agriculture in the suburbs of Asmara City, Eritrea. Int J Environ Health Res 14(1):43–52

Stark D, Al-Qassab SE, Barratt JL, Stanley K, Roberts T, Marriott D, Harkness J, Ellis JT (2011) Evaluation of multiplex tandem real-time PCR for detection of Cryptosporidium spp., Dientamoeba fragilis, Entamoeba histolytica, and Giardia intestinalis in clinical stool samples. J Clin Microbiol 49(1):257–262. doi:10.1128/JCM.01796-10

Stensvold CR, Lebbad M, Verweij JJ (2011) The impact of genetic diversity in protozoa on molecular diagnostics. Trends Parasitol 27(2):53–58. doi:10.1016/j.pt.2010.11.005

Strand EA, Robertson LJ, Hanevik K, Alvsvåg JO, Mørch K, Langeland N (2008) Sensitivity of a Giardia antigen test in persistent giardiasis following an extensive outbreak. Clin Microbiol Infect 14(11):1069–1071. doi:10.1111/j.1469-0691.2008.02078.x

Stuart JM, Orr HJ, Warburton FG, Jeyakanth S, Pugh C, Morris I, Sarangi J, Nichols G (2003) Risk factors for sporadic giardiasis: a case-control study in southwestern England. Emerg Infect Dis 9(2):229–233

Sundermann CA, Estridge BH (2010) Inactivation of Giardia lamblia cysts by cobalt-60 irradiation. J Parasitol 96(2):425–428. doi:10.1645/GE-2207.1

Suzuki J, Murata R, Kobayashi S, Sadamasu K, Kai A, Takeuchi T (2011) Risk of human infection with Giardia duodenalis from cats in Japan and genotyping of the isolates to assess the route of infection in cats. Parasitology 138(4):493–500. doi:10.1017/S0031182010001459

Takayanagui OM, Febrônio LH, Bergamini AM, Okino MH, Silva AA, Santiago R, Capuano DM, Oliveira MA, Takayanagui AM (2000) Monitoring of lettuce crops of Ribeirão Preto, SP, Brazil. Rev Soc Bras Med Trop 33:169–174. doi:10.1590/S0037-86822000000200002

Taniuchi M, Verweij JJ, Noor Z, Sobuz SU, Lieshout L, Petri WA Jr, Haque R, Houpt ER (2011) High throughput multiplex PCR and probe-based detection with Luminex beads for seven intestinal parasites. Am J Trop Med Hyg 84(2):332–337. doi:10.4269/ajtmh.2011.10-0461

Thurston-Enriquez JA, Watt P, Dowd SE, Enriquez R, Pepper IL, Gerba CP (2002) Detection of protozoan parasites and microsporidia in irrigation waters used for crop production. J Food Prot 65:378–382

Vuong TA, Nguyen TT, Klank LT, Phung DC, Dalsgaard A (2007) Faecal and protozoan parasite contamination of water spinach (Ipomoea aquatica) cultivated in urban wastewater in Phnom Penh, Cambodia. Trop Med Int Health 12:73–81

Warnecke M, Weir C, Vesey G (2003) Evaluation of an internal positive control for Cryptosporidium and Giardia testing in water samples. Lett Appl Microbiol 37(3):244–248

Westrell T, Schönning C, Stenström TA, Ashbolt NJ (2004) QMRA (quantitative microbial risk assessment) and HACCP (hazard analysis and critical control points) for management of pathogens in wastewater and sewage sludge treatment and reuse. Water Sci Technol 50(2):23–30

White KE, Hedberg CW, Edmonson LM, Jones DBW, Osterholm MT, MacDonald KL (1989) An outbreak of giardiasis in a nursing home with evidence for multiple modes of transmission. J Infect Dis 160:298–304

Wielinga C, Ryan U, Andrew Thompson RC, Monis P (2011) Multi-locus analysis of *Giardia duodenalis* intra-Assemblage B substitution patterns in cloned culture isolates suggests sub-Assemblage B analyses will require multi-locus genotyping with conserved and variable genes. Int J Parasitol 41(5):495–503. doi:10.1016/j.ijpara.2010.11.007

Willis JE, Greenwood S, Spears J, Davidson J, McClure C, McClure JT (2012) Ability of oysters (*Crassostrea virginica*) to harbour zoonotic parasites *Cryptosporidium parvum* and *Giardia duodenalis* during constant or limited exposures in a static tank system. Oral Presentation, IV International *Giardia & Cryptosporidum* Conference, Wellington, New Zealand. Page 126 of Abstract book

Willis JE, McClure JT, Davidson J, McClure C, Greenwood SJ (2013) Global occurrence of *Cryptosporidium* and *Giardia* in shellfish: should Canada take a closer look? Food Res Int 52:119–135

Wright AC, Danyluk MD, Otwell WS (2009) Pathogens in raw foods: what the salad bar can learn from the raw bar. Curr Opin Biotechnol 20:172–177. doi:10.1016/j.copbio.2009.03.006

Zaglool DA, Khodari YA, Othman RA, Farooq MU (2011) Prevalence of intestinal parasites and bacteria among food handlers in a tertiary care hospital. Niger Med J 2(4):266–270. doi:10.4103/0300-1652.93802